FORAGE CONSERVATION AND FEEDING

FORAGE CONSERVATION AND FEEDING

Frank Raymond
Richard Waltham

Fifth Edition

Farming Press

First published 1972
Fifth edition 1996

ISBN 0 85236 350 8

A catalogue record for this book is available from the British Library

Published by Farming Press Books
Miller Freeman Professional Ltd
Wharfedale Road, Ipswich IP1 4LG, United Kingdom

Distributed in North America
by Diamond Farm Enterprises
Box 537, Alexandria Bay, NY 13607, USA

Figure illustrations by Christopher Raymond
Cover photographs:
Front: Harvesting cut grass for silage (Westmac Ltd)
Back: Round bales of silage ready for winter (Volac International Ltd)

Cover design by Andrew Thistlethwaite
Typeset by Galleon Typesetting, Ipswich
Printed and bound in Great Britain by
Biddles Ltd, Guildford and King's Lynn

CONTENTS

	Preface to the Fifth Edition	vii
	Foreword	ix
1	INTRODUCTION	1
2	THE PRINCIPLES OF FORAGE CONSERVATION Haymaking – High-temperature drying – Silage	7
3	THE FEEDING VALUE OF CONSERVED FORAGES Digestibility – Forages at cutting – Conserved forages – Protein value – Feed intake – Conserved forages fed in mixed rations	36
4	CROPS FOR CONSERVATION Sward management – Fertiliser use – Diseases – Crops for hay and silage – Forage maize – Wholecrop cereals – High-temperature drying – Forage conservation / plant and animal conservation	59
5	MOWING AND SWATH TREATMENT Mowing equipment – Conditioning equipment – Crop loss and drying rate	90
6	HAYMAKING Bales and balers – Barn hay-drying	107
7	SILAGE-MAKING Harvesting the crop – Dry-matter content – Chop-length – Forage harvesters – Forage maize – Wholecrop cereals – Filling the silo – Big bale silage – Tower silos – Effluent loss – Care during storage	121
8	STRAW AS ANIMAL FEED Digestibility – Untreated straw – Alkali treatment	164

9 METHODS OF FEEDING 172

Hay and straw – Silage – Mechanised feeding – Big bales – Forage boxes – Mixer-feeder wagons – Tower silos

10 FEEDING CONSERVED FORAGES 189

Dairy cow feeding – Milk quality – Beef production – Forages for sheep

11 FORAGE CONSERVATION IN FARMING SYSTEMS 218

ABBREVIATIONS USED IN TEXT AND TABLES 227

INDEX 229

PREFACE TO THE FIFTH EDITION

The major changes in both the making and the feeding of conserved forages that have taken place since the last edition of this book in 1986 indicated the need for considerable revision. This we have done after discussion with many researchers, farm management advisors and practical farmers. In particular we would like to acknowledge the considerable contributions that have been made by our colleagues Geoffrey Burgess, of Wiltshire, and Peter Redman, of ADAS, Silsoe, to Chapters 5, 6, 7 and 9: Joe Johnson, of ADAS, Leeds, to Chapter 4; Cled Thomas, of the Scottish Agricultural Colleges, to Chapters 3 and 10; Richard Phipps, of the Centre for Dairy Research, Reading, to Chapters 4 and 10; and Gordon Newman, of Somerset, to Chapters 4, 8, and 10, as well as the valuable advice of the Kingshay Farming Trust, of Henley Manor, Somerset, and of Mike Lemmey, of Liberty Farm, Halstock, Somerset.

September 1996

FRANK RAYMOND
RICHARD WALTHAM

FOREWORD

Grassland is the basic and potentially the most cost effective source of nutrients for ruminant livestock. It is therefore surprising that only around 25 per cent of the annual productive potential of our grassland is utilised in practice. Further exploitation of this potential is vital to the future competitive strengths of ruminant livestock enterprises.

In achieving this objective forage conservation has a major role. Fortunately recent Research and Development in this area has been particularly productive in providing the platform for progress in the field. Forage plant growth and development is now more fully understood. More is known about the fermentation process of ensilage, and the range of additives available has been extended to those containing bacteria and enzymes. At the same time there have been major advances in the knowledge of the unique digestive processes of the ruminant which allow the more effective combination of forages and other feeds in the total diet of the animal.

Mastery of both the science and practice of these factors is essential to all who share the common objective of further strengthening the commercial success of our ruminant livestock enterprises.

GORDON R. DICKSON BSc, PhD, FRAgrS, FIAgM
Professor of Agriculture
University of Newcastle upon Tyne

FORAGE CONSERVATION AND FEEDING

CHAPTER 1
INTRODUCTION

Forage Conservation and Feeding was first published in 1972, just before the United Kingdom joined the European Economic Community, now the European Union (EU). At that time it was still necessary to *argue* the case for better forage conservation, for until then agricultural policy in the UK had kept the costs of cereals and animal compound feeds, relative to the returns from milk and beef, low enough for them to be fed profitably for the whole of the 'production' ration. This meant that conserved forage – mainly hay – was expected to provide only 'maintenance' feed. Thus most dairy cows were still fed 0.4 kg of concentrates for every litre of milk they produced, winter and summer, and large amounts of cereals were fed to beef cattle – to the extent that cereals made up almost the whole ration fed in 'barley-beef' production.

As a result it was not really necessary for livestock farmers to manage their grassland efficiently. Many keen farmers did, of course, seek to grow more grass, to improve their grazing, and to make better hay and silage. But in many cases their management problems would have been easier – and they would have made more money – it they had kept more animals by increasing their summer stocking rates, conserved less forage for winter feeding, and purchased more cereals and concentrates.

It is true that, by the end of the 1960s, there were signs that the output of some farm products in the UK, as well as in the EU, was beginning to get ahead of market demand, and measures such as 'standard quantities' were introduced in the UK to put some check on farm output. But before these had time to have any effect a quite new situation opened up with the UK's entry into the EU – the impact of which was even greater because it coincided with two other events in 1973 and 1974: the sudden increase in energy costs following the first 'oil crisis', and the unexpected world shortage of cereal and protein feeds resulting from a series of poor harvests in the main exporting countries. Suddenly the priority of agricultural policy, in the UK as well as in the EU, was once again to raise farm output so as to prevent a recurrence of the damaging food shortages of the immediate post-war years.

To this end farmers were actively encouraged to adopt the new high-output production systems that were just beginning to emerge from the huge agricultural research and development programmes that had been set up after the war. The results were spectacular. In the UK dairy industry the application of selective breeding, improved health and better feeding meant that the average milk yield per cow suddenly began to increase by 4 per cent a year, twice as fast as in the previous 20 years (Figure 1.1). Similarly, as a result of the combination of higher-yielding varieties, higher fertiliser inputs, and improved herbicides and fungicides, average wheat yields in the UK increased from 4 tonnes per hectare in 1972 to over 7 tonnes per hectare in 1984.

Higher concentrate feeding did certainly contribute to the rapid rise in milk yields after 1972, but so also did improved grazing and forage conservation, because the higher cereal prices within the EU now made it more profitable for dairy farmers to adopt the more efficient grassland management systems that previously

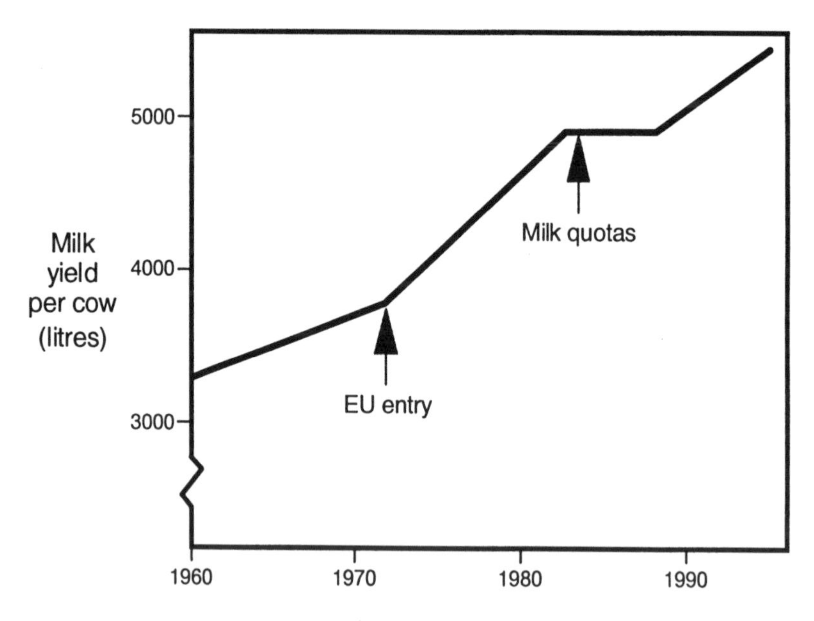

Figure 1.1 Average annual milk yield of dairy cows in the United Kingdom, 1960–93, with dates of EU entry and of introduction of milk quotas.
(Data: *Agriculture in the United Kingdom* (MAFF/HMSO))

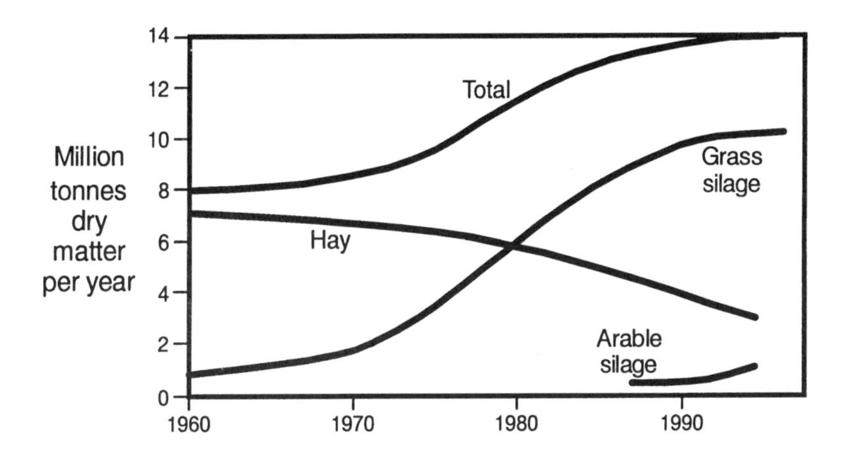

Figure 1.2 Quantities of hay and silage conserved each year in the United Kingdom, 1960–93. (N.B. The data are not exact as they are based on on-farm estimates of hay and silage production.) (Data: *Agriculture in the United Kingdom* (MAFF/HMSO))

only the grassland enthusiasts had been using. One of the most notable developments was in the shift from hay to silage as the main method of forage conservation (Figure 1.2). In 1972 very little silage was being made in the UK; in contrast, by the end of the 1980s some four-fifths of the forage conserved was as silage, with the most marked shift being on dairy farms, on which nearly half the fields were cut at least once a year for silage. Equally as important, by 1990 nearly twice as much forage was being conserved each year as had been conserved in the early 1970s.

It is now quite clear that the officials who planned the policies aimed at raising food production in the early 1970s greatly underestimated the potential of these new technologies, and so the rate at which farm output would increase. As a result, by the early 1980s the EU was faced with a huge problem of surplus food production, with intervention stores holding increasing quantities of 'unsaleable' cereals, butter, skim-milk powder and beef.

The classical solution to such overproduction had been to reduce prices, and to this end the 'real' prices (that is, allowing for inflation) paid to EU farmers for cereals, milk and beef were steadily reduced. Despite this, output continued to rise – both because the continuing stream of new technology allowed farm output to be increased still further so as to compensate for lower unit returns, and because intervention buying offered a virtually unlimited market for this additional output. It became evident that price cuts alone, at least at any politically acceptable level, would not be enough to curb excess production.

The first commodity to be tackled was milk. Milk quotas were imposed in 1984, just before the previous edition of this book was published; we were thus able to speculate on the likely consequences of this new regime, under which the individual dairy farmer could no longer compensate for price cuts by producing and selling more milk. In particular we emphasised that this indicated a new priority, to produce each litre of milk permitted by the quota as cheaply as possible, rather than to produce more milk. To this end we suggested that most farmers should aim to increase still further their reliance on forages, both grazed and conserved, and to feed less concentrates, even if this meant checking, for a time at least, the increase in milk yield per cow that had been a dominant feature of dairy farming during the previous 30 years. We were able to record that in 1984–5, the first full year of milk quotas, concentrate feeding had fallen, from 0.33 kg to 0.25 kg per litre of milk produced in the UK; yet, despite a fall of 100 litres in average milk yield, margins per hectare had been higher than in 1983–4, the last year of unrestricted milk production. Improved forage conservation had been one of the factors that had made this possible.

Ten years on there have, of course, been many further changes. The relationship of the milk price to feed prices is now higher than in 1984, so that it is again more profitable to feed concentrates; quota transfer has allowed some dairy farmers to aim for higher yields, so that average milk yields are once again rising as they did in the 1970s (see Figure 1.1). However, under a quota regime it is still essential to produce milk as cheaply as possible, so that forages, and in particular conserved forages, continue to contribute more to dairy rations than before quotas were introduced.

The following chapters examine some of the technical developments, many of them initiated by farmers, that have made this possible – for example, in the use of 'buffer' feeding of silage when there is a shortage of herbage for grazing (a technique almost unheard of even ten years ago), and in the huge increase in the production and feeding of silages made from forage maize and wholecrop cereals, shown in Figure 1.2. This greater reliance on forages, to replace some of the concentrates that were previously fed, has demanded both a higher standard of herd management and a greater attention to detail in growing and utilising forage crops. As we have noted, many of the techniques for doing this were already to hand; it needed the sanction of milk quotas to persuade more dairy farmers to adopt them – and to continue to improve them.

For a number of reasons progress in beef and sheep husbandry has been less marked, at least in part because, until very recently, there have not been the same restraints on meat production as with milk. The feeding of cattle and sheep has also tended to be more empirical, and so less well organised to benefit from technical progress, while the traditionally fragmented markets for beef and lamb have often placed almost as much importance on 'dealing' as on 'feeding'. This situation is changing, in particular as a result of the increasing power of the supermarkets in dictating the specification of the meat that must be produced. However, as we discuss in Chapter 11, the precise future contribution of forages, both grazed and conserved, to the production of that meat is still uncertain. Will the new market demand be for grain-fed meat? Will this be encouraged if grain prices fall as a result of reform of the Common Agricultural Policy? Or will the UK be able to continue to exploit its natural advantages in growing high-quality forages for meat production?

Certainly, though, the future role of forages in both milk and meat production will continue to depend critically on how efficiently they are grown and used – and in particular on how effectively the forage that is 'surplus' to current grazing requirements is stored for later feeding at times of shortage. In 1972 we argued that developments in forage conservation were being held back because hay and silage were looked on as low-value 'maintenance' feed. This meant that the methods used for making such feeds had to be cheap and simple; yet the use of such cheap methods almost guaranteed that the feeds produced *would be of low feeding value*. Thus we argued that, at least initially, research on forage conservation should be primarily concerned with biological efficiency – how to make high-value feeds with low losses – rather than with costs. We suggested that the improved conservation methods resulting from this research would not necessarily be either more complex or more expensive – but that they were most unlikely to be developed if the first consideration was cheapness.

Fortunately research into more efficient methods of forage conservation had been stepped up in the late 1960s. This meant that, when the economic pressure to make better use of forages came with UK entry into the EU, livestock farmers were quickly able to put the results of this research into practice – as seen, for example, in the rapid shift from hay to silage, noted in Figure 1.2 – and in many cases to provide the key management input needed to make the effective transition from 'the laboratory' to 'the field' that had been lacking in much of the earlier research.

Thus much of what we wrote about in 1972 is now accepted, and widely applied, on livestock farms. But there remains much scope for further improvement, in particular in raising still further the animal production potential of conserved forages, and in making conservation methods more independent of the weather, and so more predictable. To achieve this, we believe, requires first a good understanding of some of the principles which determine the feeding value of forages, the changes that occur during different conservation processes, and how these may affect the feeding value of the product; and second, how this information can be used to develop more efficient and profitable conservation systems that can be applied under practical farm conditions.

As we observed in 1972, 'It is not necessary to be a biochemist to make good silage.' But we believe, as then, that there is still great advantage in the farmer who plans to make hay or silage knowing something about the process he or she will be using, because success will so often depend critically on the informed attention to detail that such understanding makes possible. Thus the following chapters describe some of the principles underlying a range of different ways of conserving forage crops. They also describe how these are likely to relate to the particular needs of the individual livestock farm, for that is where the key management decisions will have to be taken.

CHAPTER 2

THE PRINCIPLES OF FORAGE CONSERVATION

Forage conservation is the process of cutting an actively growing forage crop and treating it so that it can be stored for feeding to livestock at a later time. Until very recently, as we have noted, this was a pretty inefficient and unpredictable process. Losses, both in the field and in store, were high; more than a third of the crop could be lost, and in a difficult year mouldy hay would be burnt in the field because it was not worth baling. The conserved product was also often of low nutritive value. However, there had been little incentive to make forage conservation a more effective process – in part because alternative cereal-based feeds were cheaply and readily available, but mainly because to conserve fresh green crops consistently well under practical farm conditions was, and continues to be, a process demanding skill and attention to detail.

Very soon after a green forage is cut the walls of the plant cells become permeable and the cells start to lose moisture, the soluble carbohydrates (sugars) in the cell sap begin to oxidise, and the proteins begin to break down. At the same time bacteria and moulds, which are always present on the surface of growing plants, but which can do little damage as long as the plants are living, begin to attack and decompose the dying tissues. The aim of an efficient conservation system is to check these destructive processes rapidly and completely, so as to preserve as much as possible of the yield and feeding value present in the crop at the time it was cut.

Two main processes are used to do this: either the crop is dried at low temperature, by haymaking in the field, or at high temperature, in a drum drier, down to a level of moisture content at which both chemical and microbial activity cease; or it is preserved, at a higher moisture content, by the action of acids or other chemicals, in the process of ensilage. Table 2.1 shows the approximate levels to which the moisture content of the cut crop must be reduced

Table 2.1 Moisture content at which the cut crop can be removed from the field for different conservation systems

Treatment	*Moisture content limits (% dry-matter content)*	*Field exposure time*
Barn-drying of baled or chopped hay, using some heated air	45–60	8–72 hours
Storage conditioning of baled or chopped hay, without supplementary heat	30–45	48–96 hours
Baled hay treated with a chemical additive	25–35	48–120 hours
Baling followed by field conditioning and barn storage of hay; mature crops	20–30	48 hours, up to 2 weeks or longer
Baling followed by field conditioning and barn storage of hay; immature crops	15–20	60 hours, up to 2 weeks or longer
Silage stored in a clamp or bunker silo	65–83 (17–35) (25 per cent target)	24–48 hours
Silage stored in sealed big bales	70–80 (20–30)	24–48 hours
Silage stored in a tower silo	60–70 (30–40)	24–72 hours

before it is removed from the field to be conserved in these different forms.

Some background understanding of these different processes is important in deciding which method of conservation is likely to be the most suited to each particular farm, and in ensuring that the method that is adopted is applied as successfully as possible.

HAYMAKING

Haymaking aims to make a dried product, generally from grass and often from a crop of only medium feeding value, which can be stored without wastage for a long period, and at a reasonable capital and labour cost. In order to do this between 70 and 90 per cent of the moisture that was present in the crop at the time it was cut has to be removed, by the action of sun and wind as the cut crop lies on the surface of the field, before it is fit to be lifted and removed for storage. The exact moisture content at which each particular crop can be safely stored depends very much on its composition and its stage of maturity at the time it is cut. Thus mature crops, which contain relatively low amounts of sugars that can oxidise and cause heating in storage, are generally safe to store when their moisture content has been reduced to between 15 and 18 per cent. In contrast hay that is made from immature, leafy crops contains a higher sugar content, and has to be dried to below 12 per cent moisture before it is safe to store without risk of heating and moulding.

Immature crops not only have to be dried down to a lower moisture content but they also generally contain more moisture than mature crops; this means that much more water has to be removed to ensure safe storage of hay made from an immature than from a mature crop. Thus 3.5 tonnes of water per ha have to be evaporated from an immature crop at 80 per cent moisture to produce 1 tonne of 'dry' hay at 12 per cent moisture content, compared with less than 2.5 tonnes of water when a more mature crop, containing only 75 per cent of moisture, is dried down to the 15 per cent at which it can be safely stored. Immature crops are also in general more difficult to dry in the field, because they contain more leafy material which packs together and reduces the movement of drying air through the crop.

On the other hand, while a mature crop is thus easier to make into hay than an immature crop, and while losses are likely to be lower, the hay produced will be of lower feeding value. Thus in deciding when each crop should be cut and which haymaking system should be adopted a balance has to be struck between the ease and certainty of the making process and the likely feeding value of the hay that will be produced.

Field-dried hay is almost always baled before it is removed from the field, and it is generally necessary to reduce the moisture content of the cut crop to below 25 per cent before it is baled. However, this is still too high for safe storage, and to produce 1 tonne of finished hay at, say, 15 per cent moisture content a further 150 kg of moisture has to be removed, out of the 3 tonnes or so present in the crop at the time it was cut. With most crops this remaining moisture is most effectively removed by building the bales of hay into stacks in the field, small enough to allow air movement through the stack so as to complete the drying process and to prevent heating. With hay made from very young crops, or with hay baled at a higher moisture content, on the other hand, the rate of moisture loss from field stacks may not be enough to complete the drying process. To prevent further heating and loss of nutritive value it may then be advisable to treat this hay with a chemical additive before it is baled. Alternatively the residual moisture can be removed by artificial ventilation in a system of barn-drying of hay; however, as discussed below, this practice is now seldom used.

The speed of field-drying depends greatly on the size of the crop that is cut – heavier crops contain more water per hectare and take longer to dry than light crops. With most commercial hay crops between 15 and 20 tonnes of water have to be evaporated from each hectare of crop that is cut. Given efficient treatment and reasonable weather, three-quarters of this water can be lost on the day that the crop is cut, because moisture readily evaporates from the outer surfaces of cut plants, down to a moisture content of about 65 per cent. However, to get the highest rate of water loss the swath of mown crop must be opened up immediately after it is cut to allow drying air to penetrate right through the crop. If this is not done only the surface layer of the swath will dry, leading to a marked difference in moisture content between the surface, exposed to sun and drying air, and the bottom of the swath, which may still be lying on damp soil. In fact, under poor drying conditions the average moisture content of an undisturbed swath may even increase because moisture is picked up from the soil, as well as being produced within the swath as a result of oxidation of the 'sugars' in the cut crop.

Thus immediate action, discussed in detail in Chapter 5, must be taken to open up the swath as soon as possible after cutting so as to speed up the rate of moisture loss. The initial rate of moisture loss from the leaves and the outer surfaces and cut ends of the stems is rapid, but once the moisture content of the crop falls to 60–65 per cent the next stage of drying, down to about 30 per cent, is much slower, because moisture now has to diffuse out from the inside of stems and larger plant fractions before it can evaporate. The rate and evenness of drying during this stage are greatly increased if the crop has been subjected, immediately after it is cut, to some form of mechanical 'conditioning' which damages the plant surfaces so as to make them less resistant to water movement (p. 96).

The final stages of drying, from about 30 per cent moisture content down to the required storage level, are greatly influenced by the prevailing weather conditions, in particular the rainfall and air humidity. Except under the most favourable conditions drying can be very slow and, particularly if rain sets in, the resulting alternate wetting and drying of the crop can lead to a progressive reduction in both yield and quality – and in extreme cases to complete loss of the crop. At this stage the physical loss of small plant fragments, in particular leaves breaking off, may be accentuated if further mechanical treatment is applied to speed up the drying process, and tedding of the crop must be done with extreme care.

Whatever the weather conditions, then, the final drying of hay down to a safe storage moisture content in the field does carry some risk of losses – which is one reason why haymaking has progressively been replaced by silage.

As has been noted, tedding and conditioning will greatly speed up the rate of moisture loss from the cut crop, but they cannot prevent the loss of nutrients that are leached out of the crop by rainfall. In fact a conditioned crop, in which the plant structure has been opened up, may well suffer greater losses as a result of repeated wetting than a similar crop which has not been conditioned. Thus efficient haymaking demands not just the right machines, but also skill in deciding how and when to use each machine in the light of the likely prevailing weather conditions.

Under poor drying conditions the combined respiration, physical and leaching losses during field haymaking can be as much as a third of the original dry matter in the crop. If the hay is not fully dried before it is stored there can then be further losses, both of dry matter and of nutritive value, from the hay heating up as a result of continued respiration and microbial activity. Although the rate of respiration does decrease as the temperature in a stack of hay rises above 32°C, respiration in moist hay continues at least until the plant cells begin to die at about 45°C. At that temperature other bacteria and moulds, the thermophilic organisms, then begin to multiply and cause further chemical decomposition and, unless these are controlled, the temperature continues to rise. If the temperature in the hay reaches about 70°C chemical oxidation begins, and there can then be a rapid rise in temperature to over 200°C, with risk of the stack catching fire.

Fortunately this degree of heating does not often occur, and the most common problems are the production either of mouldy hay, which is unpalatable and which can cause health hazards to both animals and man, or of brown, overheated hay which, though palatable, is generally of low feeding value.

In practice, though, haymaking has now largely been replaced by silage, particularly for the heaviest cuts of forage taken in May and June. This means that much of the hay that is now produced is made from second or later cuts of forage, which tend to be from lighter crops, harvested in mid-summer when drying conditions are the most favourable. Both these factors greatly improve the chances of making well-preserved hay with relatively low losses. However, haymaking in the British climate still remains a somewhat unreliable operation; in particular the decision when to bale and remove hay from the field often becomes a compromise between accepting a moisture content which may risk some heating in store, and leaving the crop to continue drying in the field, with the possibility of further physical and leaching losses.

Barn-drying of hay

As has been noted, much of the residual moisture in hay may be removed by 'curing' the bales of hay in small stacks on the field before they are brought in to store. However, this is still dependent on weather conditions, and a more certain way of minimising losses from heating and moulding in store is to stack the bales under cover and to finish the drying process by air ventilation; this is generally referred to as barn-drying. Table 2.1 shows that hay can be removed from the field after a shorter drying period, and at a higher moisture content, if it is then barn-dried than if drying is completed in the field, particularly if some supplementary heat is used in the barn.

It must be emphasised that when hay is baled and removed from the field at a moisture content above about 30 per cent, it is vitally important to ensure that the different parts of the windrows in the field have dried as evenly as possible before the crop is baled, because variations in moisture content between bales, or within individual bales, can cause real problems in a barn-drying operation.

Barn-drying was introduced into the UK in the early 1950s, but was adopted on only a few farms because of the low throughput and the high labour requirements involved in the double-handling of bales in the 'batch' systems then in use, which were suited to making up to 30 or so tonnes of hay a year. However, because the cut crop spends less time in the field, barn-drying did allow crops to be cut at a less mature stage, and so at a higher level of nutritive value, and with lower losses, than with conventional haymaking. Thus a considerable research programme during the 1960s showed that hay yields could be 15 per cent higher with barn-drying than with field haymaking; also, because the system reduced the loss of the more digestible fractions of the crop that can occur during the final stages of field-drying, barn-dried hay had a higher nutritive value than field-cured hay made from the same crop – though generally of lower nutritive value than the original crop, because some field losses were still unavoidable (Table 2.2).

Table 2.2 The digestibility (D-value) and crude protein content of hay made by different methods from a crop cut on three dates at increasing stages of maturity

Date of cutting (1958)	*Cut grass* *D-value*	*% CP*	*Barn-dried hay* *D-value*	*% CP*	*Field-made hay* *D-value*	*% CP*
6 June	65.5	11.1	62.5	10.8	58.2*	9.5
17 June	60.5	9.5	52.0*	8.5	45.7*	8.4
25 June	55.2	7.5	52.2*	8.6†	41.5*	7.8

* Crop affected by rain in the field.
† The high protein content of this hay has never been explained!
(Data: Shepperson, NIAE)

These potential advantages of barn-drying led to considerable developments during the 1960s, both in methods of ventilation and in a shift from batch drying to storage drying as the preferred system. However even at its peak barn-drying accounted for no more than a small proportion of the total hay crop made in the UK and, as field haymaking declined from 1970 onwards, it was replaced, not by barn-drying of hay, but by ensilage. Many live-stock farmers were now able to make 1,000 tonnes of silage within one week, at a lower cost in both capital and labour than storing the equivalent 250 tonnes of hay in a barn-drying installation. Thus few large-scale units were installed, and most of the limited number of barn-drying installations still operating are likely to be storage driers, holding between 50 and 100 tonnes of hay, and on farms on which most of the forage that is conserved is probably stored as silage. However, the relatively small amounts of barn-dried hay that are being made will be of very high quality, of particular value for feeding young stock and high-yielding dairy cows – and possibly in the important market for horses. (Sainfoin hay continues to have a remarkable reputation as a feed for racehorses!)

But the present, and likely future scale of barn-drying is very limited, and we have therefore omitted from this new edition of

Forage Conservation and Feeding the detailed description of barn-drying systems that were included in previous editions.

The use of chemical additives in haymaking

Over the last 20 years there has also been some interest in the possible use of chemicals, as an alternative to barn-drying, to prevent heating and moulding in hay that has been baled above a 'safe' moisture level. Under experimental conditions a number of chemicals, generally containing propionic acid as the main ingredient, have been found to prevent the development of moulding in hay containing up to 35 per cent of moisture. These chemicals act both by stopping plant respiration, which uses up some of the residual sugars in the crop and also produces further heat and water, and by controlling mould growth. Propionic acid also has a specific activity in controlling the thermophilic moulds, noted above, which begin to proliferate in hay at temperatures above 45°C; these moulds can produce toxic spores which, if inhaled from dusty hay, can cause Farmer's Lung in humans and mycotic abortion in cattle.

The initial experimental work with hay additives was followed up by field trials which showed that, to prevent moulding under farm conditions, about 6 kg of propionic acid must be retained in each tonne of hay at 30 per cent moisture content. However, because propionic acid is volatile, and because some of the acid may be lost as it is mixed with the hay during the baling process, the application rate of propionic acid may in practice need to be doubled, to 12 kg per tonne of moist hay, to ensure safe storage. An alternative chemical, ammonium propionate, is less volatile than propionic acid so that application losses are lower; it is also more pleasant to use. However, it contains only 65 per cent of the activity of propionic acid so that proportionately more has to be applied – about 6 kg per tonne of hay at 25 per cent moisture content, increasing to 12 kg at 30 per cent.

These rates have proved effective with both conventional bales and large rectangular bales, but when hay is stored in large round bales (p. 115) a higher level of additive – up to 18 kg of ammonium propionate per tonne with hay at 25 per cent moisture content – is generally advised. With round bales of hay at higher moisture contents it has proved in practice difficult to prevent moulding regardless of the amount of additive applied.

A problem in the use of hay additives has been in applying the liquid containing the additive uniformly to the hay as it is baled.

Investigation of a number of different application systems has shown the advantage of spraying the chemical solution in large droplets (700–1,200 mμ diameter) at, or very close to, the baler pick-up. These large drops minimise the loss of volatile acid, and penetratation of acid into the mass of the crop is aided by the drops shattering on impact and mixing with the crop as it passes through the baler chamber.

For effective and economic treatment the rate at which the chemical is applied must be closely related to the moisture content of the crop and to the baler throughput, both of which can vary considerably through the day. Thus the application rate may need to be altered at intervals to take account of changes in crop condition and baling rate.

While hay has lost its place as the main method of forage conservation, greater importance is probably now given, with the hay that is made, to speed of making and avoidance of losses. Chemical additives therefore still have a useful role in haymaking, and a number of commercial formulations are marketed. While these differ in composition, a particular additive is most likely to be effective in allowing hay to be baled at a higher moisture content if the recommended application rate contains either propionic acid or ammonium propionate at the rates noted above.

It is difficult to quantify the practical benefits from the use of hay additives, because they were introduced at much the same time as the improved mowing and conditioning techniques, described in Chapter 5, which greatly speeded up the overall operations of haymaking and the rate of moisture loss in the field. However it is likely that, combined with faster field making, the tactical use of additives when drying conditions are unfavourable has contributed to the slow but steady improvement that there has been in the quality of hay made in the United Kingdom, with the average protein content in hay having risen by about 1 per cent since 1970.

HIGH-TEMPERATURE DRYING

Undoubtedly the most efficient method of conserving a green forage crop is by artificially drying it with hot air. Total loss of dry matter, from standing crop to dried product, can be as low as 3 per cent; furthermore, because the crop can be cut for drying at a much more immature growth stage than is practicable for either hay or silage, the nutritive value of the dried product can be much higher. High-temperaure drying is also largely independent of weather conditions.

Because of this potential a considerable programme of research on high-temperature drying ('grass-drying') began during the 1960s, and the first edition of this book, in 1972, included a major section on the subject. Largely as a result of this research, production of dried grass in the United Kingdom rose from 65,000 tonnes in 1965 to over 200,000 tonnes in 1972, and further major expansion was predicted.

However, grass-drying is based on the burning of fossil fuel, generally oil, to evaporate the water in the fresh crop, with up to 300 litres of oil being needed to produce 1 tonne of dried grass from a crop cut at 80 per cent moisture content. The series of sharp increases in the price of oil that began in 1972 obviously made grass-drying much more expensive, and greatly reduced the competitive position of dried grass as a livestock feed. As a result there was a steady fall in the amount of dried grass produced in the United Kingdom, down to the present level of about 70,000 tonnes a year. Most of this is from drying plants with an annual output of more than 5,000 tonnes a year, and the operators who have continued in production have remained competitive by wilting the cut crops in the field before bringing them to the drying plant, thus greatly reducing fuel consumption and increasing drier output. Some operators have also installed dewatering equipment, which squeezes some of the moisture from the crop before it is dried.

The situation in the United Kingdom contrasts sharply with that in a number of other EU countries, and since 1980 total EU production has more than doubled, to 4,500,000 tonnes of dried green crop a year. The main reason has been that these other countries have made much more effective use of the support for dried green crops that was introduced in 1975 as part of the Protein Feeds Scheme of the Common Agricultural Policy. Over the following years the level of aid steadily rose, to more than 100 Ecu per tonne of dried green crop – in many cases higher than the market return from the sale of the product! While Community support has recently been reduced, the high level of support throughout the 1980s encouraged massive investment in new drying plant, in particular in France, Spain and Italy, leading to the recent surge in production. Yet, despite the availability of this EU support, there was no comparable investment in the United Kingdom. At least in part this was because most of the 'continental' expansion has been in cooperative drying units, set up to dry crops harvested from large numbers of cooperating farms, and so with strong national political support. Most UK drying units, on the other hand, are privately owned and operated. As a result there has been much less investment in new drying equipment in the UK industry and, with the cost of a new drying installation with an output of 2 tonnes of dried grass per hour now well in excess of £2 million, further significant investment seems unlikely. Thus, despite the undoubted technical and biological efficiency of high-temperature drying, we expect it to play at most only a very minor role in future forage conservation in the UK.

SILAGE

Ensilage, the process of preserving wet forage crops by fermentation, was first introduced into the UK in the 1860s ('this new method from France', as the *Journal of the Royal Agricultural Society of England* described it), and since then there have been cycles of interest in the process, with enthusiasts proposing an array of different methods of silage-making which have seldom worked out in practice. Most of the earlier research concentrated on the preservation process itself, with the aim of reducing losses during making and storage, and it is only more recently that attention has been given to the feeding value of the stored product, and to the effect that different ensilage methods can have on feeding value. In fact a number of earlier silage systems *did* give very efficient preservation, but had little practical application because the silages that were produced were of only limited value as animal feed.

When a freshly cut green crop ('grass') is stacked in a heap its temperature rises as a result of the heat produced by chemical reactions that take place within the cut crop. Initially most of this heat is generated from the oxidation, by the oxygen trapped within the mass of the crop, of the water-soluble carbohydrates ('sugars') which are contained in the plant cell solution. Once heating starts it can rapidly speed up, partly because chemical reactions take place faster at higher temperatures, but mainly because the warm air produced by these reactions rises out of the heap and draws in fresh air – similar to the draught through a domestic fire. This makes more oxygen available and as a result the sugars in the crop can quickly be burnt up; yet, as described below, these sugars play a key role in the ensilage process. Thus the first priority in silage-making is to stop the oxidation of sugars by preventing fresh air getting into the cut crop. This can be done by compressing and consolidating the crop by rolling it so as to restrict air movement, with exclusion of air being much more rapid with short-chopped than with long forages. However, compression by tractor rolling can itself produce a 'bellows' action which can draw in more fresh air, and care is needed in the amount of rolling applied. In practice the most effective way of preventing air movement through the silage is to stop the hot air escaping, and so drawing in further fresh air, by covering the surface of the crop with plastic sheeting. This is the basic principle of the sealed silo systems described in Chapter 7.

However, even if the access of fresh oxygen is prevented, the grass in the heap is still completely unstable because other chemical reactions continue. Thus the proteins in the crop begin to break down to produce amino acids and ammonia; more seriously, bacteria and moulds, which are always present naturally on the grass when it is cut, can rapidly multiply and begin to decompose the crop into a putrefying and evil-smelling mass – as is clearly seen when garden lawn-mowings are left in a heap. Both chemical breakdown and undesirable microbial activity must therefore be stopped as quickly as possible; this is done either by sterilising the crop or, more generally, by making it acid.

In practice most silage systems depend on the action of acids, the most obvious solution being to add acid to the crop. This was the basis of the system developed in Finland in the 1930s by A. I. Virtanen, in which the cut crop was treated with a mixture of sulphuric and hydrochloric acids (AIV acid) before it was put into the silo; other acids, including phosphoric acid, were also used. This method, effectively 'pickling' the crop, gave very efficient preservation. However, although it was at one time fairly widely used in Scandinavia, it suffered from the very obvious disadvantages that the operator had to handle large quantities of corrosive acids and that a strongly acid effluent seeped out of the silo – and the less immediately evident disadvantage that the resulting silages were of only limited value as animal feed because their high acid content restricted the amount of them that animals were able to eat.

The principle of the AIV method was to add enough acid to the crop to produce an immediate and complete cessation of both chemical and microbial action. In contrast, most modern silage methods aim to exploit the fact that, under well-controlled conditions, most green crops undergo a natural fermentation process, with one of the main end-products being lactic acid. Thus among the range of organisms always present on the surface of green crops when they are cut are the lactobacilli, a group of micro-organisms which, in the absence of oxygen (anaerobic conditions) can ferment the sugars in the crop to produce lactic acid. The lactobacilli themselves are relatively insensitive to the acid they produce; in contrast, many other 'undesirable' bacteria and moulds, which are also present on the cut crop, are much more sensitive to acid and are largely inactivated by levels of lactic acid at which lactobacilli can continue to grow and to produce further acid.

The level of acidity in the crop is measured in terms of pH; this is an index widely used by chemists, but potentially confusing to the silage-maker because a *decrease* in pH represents an *increase* in acidity – thus silage of pH 4.0 is more acid than silage of pH 5.0. A further complication is that pH is expressed on a logarithmic scale, so that it takes ten times as much acid to reduce the pH of a crop from 5.8 down to 4.8 than from 6.8 (the normal, very slight acidity of the fresh crop – pH 7.0 is neutral) down to 5.8 – and one hundred times as much acid for a reduction in pH down to 3.8. This is important, because it means that the *initial* reduction in pH, down to about 5.0, which is needed to give some control of the undesirable micro-organisms, does not require a lot of lactic acid to be produced; but the lactobacilli will only produce this acid if conditions within the crop are anaerobic – hence the vital importance of preventing air getting into the silo right from the start of the silage-making process. Lactic fermentation also gets under way much more rapidly with chopped than with long crops, because chopping breaks the cell walls of the crop and makes the sugars dissolved in the cell solution more immediately available for fermentation by the lactobacilli.

Although the rates of both protein breakdown and undesirable microbial activity in the cut crop are greatly reduced by a fall in pH to 5.0 they do still continue, and a further reduction in pH, in many cases to below 4.0, is often needed to stop this activity completely, so as to ensure safe long-term storage. To bring this about more lactic acid is needed. However, the rate at which further lactic acid is produced, and so the rate at which the pH falls, now begins to slow down (a) because less sugar is now available to be fermented to acid than was present in the original crop, (b) because, although the lactobacilli are still able to ferment sugars to lactic acid, their rate of activity *does* decrease as the pH of the crop falls and (c) because, as noted above, progressively more acid is needed to bring about each further 0.1 unit decrease in pH.

Many crops, in particular very wet crops, do not contain enough fermentable sugar to give the reduction in pH needed to produce a stable silage. The reason is that the higher the moisture content of the crop that is being ensiled the greater is the concentration of acid – and so the lower is the pH – needed to prevent further chemical and microbial activity. *This means that more acid has to be produced to ensile wet crops than dry crops;* yet it is precisely such crops – leafy grass, well fertilised with nitrogen; grass/clover mixtures; and autumn grass – that are the most likely to contain low amounts of the sugars that the lactobacilli need to ferment into lactic acid. A further problem is that, unless such crops are rapidly acidified, the natural enzymes present in the plant cells can break down the proteins in the plant sap into amino acids and ammonia, which neutralise some of the acidifying effect of the lactic acid by a process termed 'buffering'. This reinforces the importance of the rapid establishment of anaerobic conditions, so that the lactobacilli can ferment the sugars in the cut crop to produce the lactic acid needed to prevent these chemical changes.

However, even when, by a combination of consolidation and careful sealing, anaerobic conditions are established, the amount of fermentable sugar in the crop may still not produce enough lactic acid to reduce the pH in the silage down to the level needed for stable preservation. Under these conditions the activity of damaging micro-organisms may only be checked, and they may continue to degrade the proteins in the crop. More seriously, they may begin to decompose some of the lactic acid on which the silage preservation depends. Once this occurs the pH of the silage starts to rise, and the damaging process of secondary fermentation sets in, producing the evil-smelling silage disliked by both animals and farm workers (not to mention their wives), with its characteristic butyric acid (rancid butter) and ammonia odour.

This was the recurring practical experience with silage until the late 1960s, and explains why, after a hundred years, uptake of silage-making remained at such a low level (see Figure 1.2). It thus became evident, in order to make good silage consistently from often difficult crops, that other measures, in addition to efficient sealing of the silo, were needed.

Wilting

Problem crops will not make good silage because, even if all the sugar they do contain is fermented to lactic acid, the amount of acid produced will not be enough to reduce the pH to a stable level. This is generally because these crops are of such high moisture content that very large amounts of acid are needed to reach this pH. The first step, then, must be to reduce the moisture content of the crop before it is put into the silo, for much more sugar has to be fermented to lactic acid to reduce the pH of a direct-cut crop (say 18 per cent dry matter content) down to the level of 3.8 necessary for long-term preservation than to reduce the pH of a crop wilted to, say, 25 per cent dry matter down to the level of 4.3 at which it will store safely. (The convention should be noted that the condition of crops for haymaking is generally described in terms of their moisture content, but for silage in terms of their dry-matter content; thus a crop for hay would be described as having 80 per cent moisture content, but the same crop, for silage, as having 20 per cent dry-matter content; see Table 2.1.)

The process of field wilting before the crop is brought in to the silo is described in detail in Chapter 5, and the two operations of mowing followed by wilting have now largely replaced the direct-cutting operation which predominated before 1970. Wilting has many advantages: it reduces the weight of crop that has to be carted from the field and loaded into the silo; it greatly improves the ensilage process and the feeding value of the silage that is made (p. 50); equally important, wilting greatly reduces the loss of acid effluent from the silo, which is produced in huge amounts when wet crops are ensiled and which can pose a serious environmental hazard (p. 159).

Wilting is, of course, essential if the crop is to be stored in a tower silo (p. 156), for which crop dry-matter levels above 30 per cent are required in order to reduce the pressure (and risk of damage) on the walls of the silo and the amount of effluent flowing from the silo. At this level of dry matter only a moderate degree of acid fermentation is needed to reduce the pH to between 4.5 and 5.0, at which the silage should store safely under the airtight conditions within the tower silo. With an efficient field operation and favourable weather, wilting the crop to between 30 and 35 per cent dry matter should be possible within 24 hours of mowing.

However, tower silage does still remain a weather-dependent system, because field wilting is essential to the operation and, under unfavourable weather conditions, the crop may suffer significant losses in the field before it is dry enough to load safely. In contrast the degree of wilting generally aimed at for crops that are to be ensiled in bunker or clamp silos is more commonly around 25 per cent dry matter, which is much more reliably achieved than the 30 per cent plus needed for tower silos.

Wilting even to 25 per cent may not always be practicable, though. Wet crops are often wet because the weather is unfavourable for wilting; crops cut early and late in the season have high moisture contents and, even when the weather is fine, drying rates tend to be slow; and, particularly on smaller farms, the separate operations of cutting, wilting and lifting may not always be possible with the available labour.

Thus, while the silage-maker should plan to wilt each crop before it is ensiled, he also needs a back-up to allow silage-making to continue even when the condition of the crop (low dry-matter content and low sugar content) or the weather are unfavourable. The following paragraphs thus examine the role of a number of agents – including chemicals, microbial inoculants and enzymes – that have been introduced to make the ensiling of green crops a more reliable process.

Chemical additives

The commonest cause of poorly preserved silage is that the crop, even if it is wilted, contains too little water-soluble carbohydrate (sugar) to produce the required amount of lactic acid. In that case it should be possible to add enough extra sugar to the crop to ensure a stable acid fermentation. In fact molasses, applied at about 5 kg per tonne of fresh crop, increasing to 15 kg with very wet or clovery crops, has been used as a silage additive for many years. Yet success in practice has been very variable, in part at least because of the difficulty of mixing the molasses uniformly with the crop before it is ensiled. (Molasses, even when diluted, is viscous. One author's first experience of silage was of lucerne being buck-raked on to a clamp silo. Diluted molasses was spread at intervals from a watering-can with the 'rose' removed because it slowed down the process!) Non-uniform mixing produces patches of well-preserved silage interspersed with patches of poorly preserved (high pH) material; during storage the putrefaction from those patches can spread and ruin the quality of much of the remaining silage.

It is in practice most difficult to spread molasses uniformly by hand, and spreading is most effectively done by fitting a container for molasses either directly on to the forage harvester, or on the towing tractor, from which it is delivered by gravity feed directly on to the swath or windrow of the crop as it is picked up. It then mixes with the crop as it passes through the harvester (this system was demonstrated to members of the British Grassland Society on their visit to Ulster in 1959, but few of us then recognised its significance). The relatively large amounts of molasses that are needed, and its slow flow properties, also demand an efficient handling system in order to avoid delays in the overall harvesting process. For this reason some farmers have fitted a tractor-mounted applicator to 'dribble' diluted molasses on to the crop while it is being loaded and consolidated in the silo, although this is likely to give less uniform application than addition at the time of harvesting.

Silage-making is a chemical process. It should therefore be possible to control the process by the addition of chemicals, and this possibility has been studied by many chemists over many years. The use of strong mineral acids to give rapid 'pickling' of the crop in the AIV process has already been noted. However, most research has aimed at *reinforcing* the natural fermentation process, rather than relying solely on preservation by the applied chemical, as with AIV acid.

Yet despite a considerable research effort, until the mid-1960s the use of chemical additives had led to virtually no improvement in practical silage-making – and for much the same reason that molasses had failed. For very little of this research (much of it carried out in mini-silos in the laboratory, far removed from silage-making on the farm) had recognised the key importance of mixing the additive uniformly with the crop. This was particularly the case with a number of commercial additives, including sodium metabisulphite, which were applied in powder form, but liquid additives had also given little practical benefit. The situation was transformed in the mid-1960s by the development, by Naerland in Norway, of an applicator which fed the chemical additive directly into the cutting mechanism of the forage harvester (Plate 2.1). Using this type of applicator, a number of chemicals, which had been shown to be effective in the laboratory but which had had only limited success in the field, were found markedly to improve silage-making on the farm scale. Backed up by an expanded research and development effort by both state and commercial research centres, the use of chemical additives rapidly increased, and this provided one of the key factors leading to the huge increase in silage-making that took place in the 1970s (see Figure 1.2).

The most widely used additives have been those based on formic acid, used either alone or in mixture with other chemicals. This acid had been studied in Germany as long ago as 1923; but without an effective applicator formic acid had had little significant impact on practical silage-making. Following the Norwegian experience,

Plate 2.1 Silage additive from the two 25 litre containers is fed by gravity into the cutting cylinder

research on formic acid started in the UK in 1967; this showed that the application of 1.7 kg of a liquid additive, containing 80 per cent of formic acid, to each tonne of fresh crop, rising to 4.5 kg of additive with very wet or clovery crops, gave a consistently improved preservation compared with similar crops ensiled without the additive.

Formic acid applied at these rates gives an immediate and rapid fall in the pH of the crop, to about 5.0. As noted on p. 19, this level of pH is not low enough to give long-term preservation; however, provided air is excluded, conditions are now favourable for the lactobacilli present on the crop to continue to ferment the sugars in

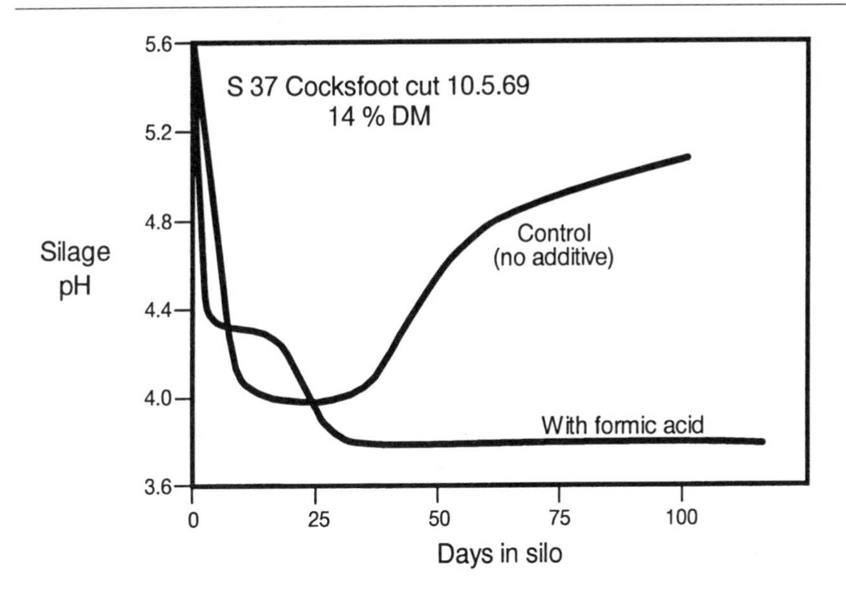

Figure 2.1 Effective silage additives (for example formic acid) prevent the secondary (clostridial) fermentation which can follow the slower initial fall in pH when a 'problem' crop is ensiled without an additive

the plant cells to produce lactic acid, leading to a further fall in pH. In many cases the final pH of the silage is only slightly lower than it would have been if the silage had been made without formic acid. However, as Figure 2.1 shows, with no additive there is a time-lag before the natural fermentation process gets under way, and as a result the initial fall in pH is slower; this allows some protein breakdown to occur and, through 'buffering' action, this can reduce the long-term stability of the silage. Thus with 'difficult' crops, ensiled without an additive, although the pH of the silage after several weeks may appear adequate, it may not be low enough to prevent the onset of secondary fermentation, leading to a progressive deterioration in silage quality.

After detailed trials on research stations and husbandry farms, followed by more extensive trials on selected commercial farms, formic acid was marketed in 1969, and was soon being widely used. Encouraged by this success, and recognising that silage additives offered a useful new market, chemical companies introduced a large number of alternative silage additives, and by the mid-1990s more than 150 different additives were being marketed. This posed a considerable practical problem for farmers and their advisors, for the claims made for many of these additives were not backed up by adequate experimental or field-trial data. Yet the cost of setting up a comprehensive 'official' testing programme would have been quite unrealistic; furthermore, there was the risk that the formulation of an additive which 'failed' the test in one year could then be changed by the inclusion in the following year of a new 'ingredient X', so that it would then need further testing.

A most important development was the publication, in November 1995, of a summary of data, *provided by the manufacturers*, and independently assessed by a panel set up by the UK Agricultural Supply Trade Association (UKASTA) jointly with the three official advisory services, ADAS, the Scottish Colleges, and the Department of Agriculture in Northern Ireland. This publication (the Register) gave data on the active ingredients, rate of application and cost per tonne of crop ensiled; more importantly, it assessed the reliability of the manufacturers' claims in terms of the nine criteria noted in Table 2.3.

Table 2.3 The UKASTA / Advisory Services Silage Additive Approval Scheme. Additives are assessed in the following categories, on the basis of an independent assessment by ADAS, the SAC and DANI of data submitted by the manufacturers

Category	*Criteria*
C1	Improved silage fermentation
C2	Improved aerobic stability of the silage produced
C3	Reduced effluent loss from the silo
C4	Reduced overall ensiling losses
B1	Improved voluntary intake of the silage
B2	Improved digestibility of the silage
B3	Improved efficiency of utilisation of energy and/or protein in the silage
A1	Improved animal liveweight gains when the silage is fed
A2	Improved milk production when the silage is fed
	Improved milk butterfat and protein will also be taken into account

On this basis only 73 of the 106 products that had been submitted were approved; more significantly, for only 44 products had experimental data been presented to show that animal performance was improved when silage made with the product had been fed to beef or dairy cattle (Category A/1/2).

The Register now provides the silage-maker, for the first time, with reliable data on the likely effectiveness of all the main commercially available silage additives, as an essential aid in deciding which additive is likely to be the most effective and economical under his/her particular conditions. It is planned to update this Register annually, again using data provided by the manufacturers. However, the actual experimental data will not be published; more detailed information on a limited number of the most promising additives is available to subscribing members of the Kingshay Trust, which carries out large-scale animal production trials on silage additives on behalf of its members.

While the decision as to which additive to use will be greatly aided by information such as that in the Register, it will be much more secure if it is also based on an understanding of the way in which the different classes of additive operate. The following paragraphs thus examine some of the factors which determine how different silage additives function, and relate these to the criteria in the Register.

(a) Inorganic (mineral) acids

Sulphuric and hydrochloric acids were used in the original AIV process. The aim was to add enough acid to rapidly 'pickle' the crop, so that there was no need to rely on a lactic acid fermentation, but AIV silage was not readily eaten by livestock because of its high acid content, and there were safety problems with the strong acids used.

However, with a better understanding of the ensilage process, and with the availability of efficient applicators, there has been a renewed interest in the use of mineral acids, and in particular of sulphuric acid. Interestingly, the initiative came from farmers, who found that sulphuric acid, applied at much lower rates than had been used in the AIV process, rapidly reduced the pH of the crop to a level at which a dominantly lactic acid fermentation would continue – and also that, purchased as a 'bulk' chemical, this acid was cheaper than most alternative commercial silage additives. Mixtures of sulphuric acid with formic acid and propionic acid, as well as with trace elements, are also marketed. As with other acid additives, the rate of application must be adjusted to the state of the crop being ensiled, in terms of moisture content and sugar content. Thus recommended rates range from 1.5 kg of acid per tonne of fresh crop at 25 per cent dry-matter content up to 2.5 kg with crops below 20 per cent moisture content, or with crops of high clover content or harvested in the autumn. While good results have been reported from many livestock farms, to date the Register has given only limited approval to any products based on sulphuric acid, at least in part because its commercial potential is unlikely to justify the high cost of obtaining the necessary experimental data.

Great care is, of course, needed in handling sulphuric acid, both in further diluting the 40 per cent concentrated acid, which is generally the form in which it is purchased, and in loading and operating the applicator. Protective clothing and goggles are essential. The acid is also highly corrosive to metal, and all parts of the forage harvester that may be contaminated with acid must be thoroughly washed down after use. One brand claims to contain a corrosion inhibitor to reduce the risk of damage, but it is notable that few agricultural contractors are willing to use their own machines to apply this acid to their clients' silages.

(b) Organic acids

The most commonly used acid is formic acid, whose action and rates of application have already been described. Acetic, lactic and propionic acids have also been used, but in general they are less effective than formic acid. When they are applied organic acids produce an immediate fall in the pH of the crop (although less marked than when sulphuric acid is used), but further lactic acid still needs to be produced during the subsequent fermentation in order to produce a silage that will be stable during prolonged storage. Organic acids are less corrosive than mineral acids, but care must still be taken when they are used. Propionic acid is also claimed to reduce surface moulding when the silo is opened.

(c) Mixtures of organic acids with formalin

In the early 1970s there was much interest in the possible use of formalin to preserve forage by partial sterilisation, as an alternative to acidification. It was considered that formalin would also improve the feeding value of the protein in the silage by reducing the extent of protein breakdown during the ensilage process, and by chemically reacting with some of the protein to produce 'rumen-undegradable' protein (p. 45). However, it was found in practice that control of the application rate was very critical; too much formalin gave good preservation but damaged the nutritive value of the silage; too little gave poor preservation and risked secondary fermentation. Thus formalin is now used only in mixtures, generally with sulphuric or formic acid. Practical experience is that these mixtures are safer and more pleasant to use than the additives based only on the acid; of the eight products on the Register containing formalin four are reported to have given improved liveweight gains with beef cattle.

(d) Other sterilising agents

A number of other chemicals tested have had bactericidal activity, rather than the mainly bacteriostatic activity of acid additives. In addition to formalin these include sodium acrylate, sodium metabisulphite and sodium sulphite. However, rather large amounts of these chemicals are needed; they also have only limited preservative activity, so that the silage face tends to mould when the silo is opened (p. 176).

(e) Sugars

Many experiments have shown that, for an efficient lactic acid fermentation, the crop should contain a minimum of 3 per cent of water-soluble carbohydrates (sugars) in the fresh weight (15 per cent in the dry matter at 20 per cent dry-matter content). The use of molasses to reinforce the natural sugar content in crops containing less than this level has been noted; the practical effectiveness of molasses has been much improved by the development of improved application systems, and by the availability of new formulations of lower viscosity, which are easier to mix with water. However, the volume that needs to be applied, generally between 5 and 15 litres per tonne of fresh crop, is greater than with most alternative additives, and good organisation is needed if this is not to slow down the overall harvesting operation. Molasses is completely safe to handle and use, though the harvester should still be washed down daily to prevent the build-up of a sticky layer. Molasses has the further advantage that it adds some (non-crop) energy to the silage that is made; the Register records improved production by both beef and dairy cattle when 'molassed' silages have been fed.

(f) Bacterial inoculants

Even when an acid additive is used, most silage-making still depends on the fermentation of sugars present in the crop to produce the lactic acid needed to reduce the pH to a safe storage level. This fermentation is initiated by bacteria, in particular the lactobacilli, which are naturally present on the crop when it is cut. However, some crops may not provide enough bacteria, or the right types of bacteria, to initiate the rapid lactic fermentation that is needed. Thus research over many years has examined the possibility of adding extra bacteria to the crop before it is ensiled, to speed up the fermentation process. Yet until very recently this has had only limited success, because not enough was known about the optimum types of bacteria. The numbers of bacteria that were needed were also greatly underestimated, and many of the bacterial cultures that were used contained only small numbers of living (viable) organisms.

New research has now isolated more active strains of a number of bacteria, including in particular *Lactobacillus plantarum* and *Pediococcus acidilactici*. As a result there has been a surge of interest in the development of silage inoculants, and the viability and effectiveness of commercial silage inoculants has been markedly improved. In addition, a practical method has been developed of distributing freeze-dried bacteria, which are then cultured on the farm to provide a daily supply of highly active micro-organisms for application to the harvested crop. Thus 72 inoculant products were listed in the 1995 Register, of which all but four contained *L. plantarum*; most importantly, half of the approved products were reported to have given improved animal production. In all cases the aim must be to ensure that at least 1 million viable (colony-forming) bacteria are applied to each gram of fresh forage that is ensiled.

A recent analysis of the results of many experiments has shown an overall advantage from the use of an active silage inoculant, even when the dry matter and sugar contents of the 'control' crop have appeared to be fully satisfactory. In all cases the inoculated silages have had a slightly higher dry-matter content than the control silages, together with a lower loss of dry matter during the ensilage process, a higher level of voluntary intake, and higher daily gains when they have been fed to beef cattle (Table 2.4). Recent research, again at the Institute for Grassland and Environmental Research (IGER), has suggested that this may be because 'swamping' the crop with active bacteria before it is put into the silo may effectively inactivate the organisms that decompose proteins to amino acids in the early stages of the ensilage process – and that the protein that is saved may play an important role when the silage is fed (see Figure. 3.1).

(g) Chemical enzymes

To be effective bacterial inoculants still require the crop to contain fermentable sugar, and there is thus interest in the use of specific chemicals (enzymes) which have the ability to break down some of the more complex polysaccharide compounds present in the crop into sugars which can then be fermented (for example, cellulase and hemicellulase, which can break down respectively cellulose and hemicellulose, the two main components of the fibre in crop plants). To date there has been little evidence that enzymes, used alone, have improved silage fermentation (only two of the seven products listed in the Register have shown any benefit), and their main use has been in combination with a bacterial inoculant. Thus 53 of the commercial inoculants included in the Register also included enzymes, and 24 of these mixed products have gained approval in category A (animal production). However, further research is needed to establish how far enzymes may have contributed to the effectiveness of the inoculants recorded in Table 2.4.

Table 2.4 A summary of the results of experiments comparing silages made with and without a biological inoculant

Silage character	*Control, no inoculant*	*With inoculant*
% dry matter in silage	20.0	22.0
Dry-matter loss during ensilage (%)	25.0	22.5
Dry-matter intake (kg per day)	8.0	8.1
Liveweight gain (kg per day)	0.8	1.0

(Data analysed by IGER)

A particular practical advantage of silage bacterial inoculants, utilised with or without enzymes, is in their safety in use. Both for this reason, and because of the increasing (public) preference for biological rather than chemical methods of control in agriculture, their use is likely to increase: already more than two-thirds of the products 'approved' in the Register are biological, and they seem likely increasingly to dominate the 'silage additive' market.

The silage-maker thus now has available a wide range of additives which, with few exceptions, aim to reinforce the natural process of fermentation of the sugars in the crop to produce lactic acid. It must be emphasised, however, that it is not necessary to use an additive in order to make good silage; the main need for an additive is to reduce the risks when silage is made from problem crops – crops which are immature or very wet, particularly when wilting is difficult; crops with high protein or low sugar contents; and crops with high legume content or cut late in the season. These different aspects are brought together in Table 2.5, the 'star' system, developed by the old Animal and Grassland Research Institute, ICI plc and the late-lamented Liscombe EHF.

Table 2.5 Guide to the use of silage additives. The number of stars (*) relating to the particular crop and harvesting conditions are added up. Above 20* no additive should be needed; between 15* and 20* the normal recommended rate of application should be used; below 15* a higher rate is indicated

Number of stars	*****	****	***	**	*
Forage species	Italian ryegrass	Perennial ryegrass	Other grasses, or grass/clover		Mainly legumes
kg N/ha	—	—	Less than 50	50–100	More than 100
% dry matter in the crop	Over 25	—	20–25	—	Less than 20
D-value of crop	—	Less than 60	60–65	Over 65	—
Method of cutting	Precision chop	Double chop	Flail	Forage wagon	—
Season	—	—	Spring and summer	—	Autumn

(Data: AGRI, ICI plc and Liscombe EHF)

But wilting may not always be practicable – or may mean delay in cutting the crop, or leaving the cut crop in the field for an extended period, both of which are likely to reduce the feeding value of the silage that is made. Thus a particular value of silage additives is that they allow crops to be ensiled that have only been partly wilted – or in some cases not wilted at all – so that cutting and lifting can be carried out to a planned timetable, except under the most adverse weather conditions. However, effluent flow when unwilted crops are ensiled can be as high as 200 litres per tonne of fresh crop, and strict precautions must then be in hand to prevent this effluent escaping (p. 159). Alternatively, an absorbent additive can be mixed with the crop before it is ensiled. Commercial products, including sugar beet pulp, cooked cereals and caustic-treated straw have been used, at up to 50 kg per tonne of fresh crop. Of these the Register records that both molassed and unmolassed beet pulp have given increased daily gains when the resulting silages have been fed to beef cattle. This potential advantage, together with the reduction in effluent production, must be considered in deciding whether their use is economic.

There is also now much practical experience of the advantage of applying a reduced level of additive to crops that have been over-wilted, which may be difficult to avoid in very hot weather. Such crops do not readily consolidate in the silo, and so quickly begin to heat up, even with daily sealing. An acid additive rapidly inactivates the oxidising enzymes in this type of crop, so reducing the extent of heating and making more of the sugar content available for fermentation to lactic acid.

Finally, as the latest data in the Register show, many additives appear to improve the nutritive value of the silage that is made, both by increasing the amount of the silage that livestock can eat, and by changing its chemical composition compared with similar silage made without an additive. Such silages are likely to have a higher potential for animal production, of particular importance in livestock enterprises in which the aim is for silage to make up a high proportion of the total ration. This aspect will be of increasing significance in the future development of the Additive Approval Scheme.

Conservation by alkali treatment

At one time the growing of wholecrop cereals for silage was fairly commonly practised. However, the crops were harvested at a relatively immature stage and, although they generally contained enough water-soluble carbohydrates (sugars) to give a rapid lactic acid fermentation, yields were low and the cost per tonne of dry matter was higher than from well-made grass silage. Thus by the 1960s very little wholecrop silage was made.

It was recognised that yields would be much higher if the wholecrop was allowed to grow to a more mature stage before it was harvested. However, more mature crops are also drier, and practical experience had been that when they were stored in conventional silos losses were very high, particularly from wastage at the silage face during feeding out. High dry-matter wholecrop silage could be successfully stored in tower silos (p. 156), but these were installed on only a few farms.

Thus, as described in Chapter 4, research was initiated on alternative ways of conserving wholecrop cereals harvested at a more mature stage. It was already known that bacteria and moulds could be controlled as effectively by high pH (alkaline conditions) as by low pH (acid). Work at the Grassland Research Institute at Hurley then showed that wholecrop wheat and barley, harvested at a fairly mature stage, could be preserved efficiently by treatment with sodium hydroxide (caustic soda) or with aqueous or anhydrous ammonia before it was stored. There was concern, though, about the safety aspects of handling large amounts of strong alkali (caustic soda was added at 5 per cent of the dry weight of the crop) or of ammonia. Interest thus turned to urea, a common agricultural chemical, which can be safely and easily applied to the wholecrop as it is harvested, and is then converted to ammonia by an enzyme, urease, which is present naturally in the fresh, moist crop. The ammonia gas that is produced rapidly diffuses throughout the mass of the crop and raises the pH to above 8.0, at which microbial activity stops. The research showed that urea, added at 4 per cent of the dry weight of the crop, gave effective preservation of wholecrop cereals cut in the 45 to 55 per cent dry-matter range, at which yields are close to maximum (p. 82).

Urea is much less effective when applied to immature, higher-moisture crops because the amount of ammonia produced is insufficient to raise the pH of the wet crop to a level at which clostridial activity is prevented. Thus the ensilage of wholecrop cereal cut at an immature stage still depends on the establishment of an acid fermentation in the silo. At the other extreme the activity of the urease present in wholecrop harvested at a dry-matter content above 65 per cent may not produce ammonia rapidly enough to prevent moulding in the stored crop; one commercial additive thus contains both urea and the enzyme urease, which speeds up the conversion of urea to ammonia.

Under most conditions the optimum harvesting stage, in terms both of yield and of storage, is when the wholecrop cereal contains between 45 and 55 per cent dry matter. The Advisory Services and the Maize Growers' Association have now produced 'growers' guides', based on 'crop colour' (yellow, hint of green, etc.) and 'grain texture' (soft Cheddar, hard Cheddar, etc.), as the basis for the field decision when each crop should be cut. Particularly in hot, dry weather this optimum condition may last only a few days, and when it is reached wholecrop cereals must be harvested and ensiled as rapidly as possible. The crop is direct-cut, without wilting, and preferably fine-chopped to 20–40 mm length to increase trailer capacity, to ensure rapid consolidation in the silo and, particularly with more mature crops, to 'crack' a proportion of the grains to improve their nutrient availability.

Urea, at 12–20 kg per tonne of fresh crop, is applied either in solid form or in solution, as the crop is being harvested. Ammonia is quickly produced, and the crop must be loaded into the silo as rapidly as possible, and all exposed faces covered with plastic sheeting so as to prevent loss of ammonia gas from the cut crop. Careful sealing of the silo immediately after filling is completed is essential to ensure long-term storage stability (p. 159). It may also be possible to extend the period of harvesting to include somewhat drier crops by using the proprietary urea additive containing urease, noted above.

It has also been found that when wholecrop cereal is preserved with either sodium hydroxide or ammonia (urea) there is a small increase in the digestibility of the cut crop, as a result of the action of the alkali on the lignocellulose in the plant fibres (p. 83). Wholecrop treated with ammonia has a further advantage, compared with treatment with sodium hydroxide, that when the silage is fed it can provide the rumen micro-organisms with a source of readily available N compounds (rumen-degradable protein (p. 46)). However, rations containing a high proportion of urea-treated wholecrop silage may provide this form of protein faster than the rumen organisms can utilise it; in that case the excess N is nutritionally wasted and may, in some cases, be toxic. Care is thus needed to ensure that only as much urea is applied to the crop as is needed to give efficient preservation, and that when the urea-treated silage is fed it provides no more than half the total ration.

Whether hay or silage is made, the feeding value of the product will depend at least as much on the 'quality' of the crop that is cut as on the method of conservation that is used. Some of the factors determining the nutritive value of conserved forages, and the effects of different conservation methods, are considered in the next chapter.

CHAPTER 3

THE FEEDING VALUE OF CONSERVED FORAGES

Conserved forages are seldom given as the sole feed to ruminant livestock, but they can still provide energy and protein on the farm more cheaply than most alternative feeds. Thus the aim must be to include as much hay or silage in the total ration as is possible while getting good animal output.

This is not difficult with stock at low levels of production, such as store cattle and sheep, rearing heifers, and late-lactation and dry dairy cows. It is with the high-yielding dairy cow, rapidly growing beef cattle, and pregnant and lactating ewes that the problem arises of how to feed large amounts of hay or silage without restricting the level of output – and for which reliable information on the nutritive value of these feeds is so important.

Over the last 20 years there has been a marked shift from hay to silage as the main method of forage conservation, and this has been reinforced by the development of big-bale silage, which has led to much more silage being fed, particularly on beef and sheep farms. At the same time concerns about pollution from silage effluent, and about the poor animal production that often results when wet silages are fed, have led to a significant shift towards higher dry-matter silages. This has posed practical problems in getting the required level of field-wilting of forage crops, and as a result there has been renewed interest in maize and wholecrop cereals for silage, as reported in Chapter 4. Thus while the grass crop still makes up most of the forage that is cut for conservation, it is now often only one component of the total conserved forage stored for winter feeding.

This chapter examines the factors that determine the contribution that conserved forages, and in particular silage, can make to the feeding of high-producing ruminant animals. Under most practical feeding situations an animal's level of production is determined mainly by the amount of feed that it eats (intake), and by how efficiently it digests and utilises that feed. Thus in the case of a conserved forage we need to understand the factors that control how much of that forage animals can eat, and how efficiently they digest the forage that is eaten. Then, because the forage will seldom be given as the only feed, but is likely to be fed in a mixed ration, practical feeding requires a knowledge of the way in which it is likely to interact with the other components of the ration.

DIGESTIBILITY

The most important single factor determining the nutritive value of a feed is its digestibility , which is most most usefully described by the equation:

$$\text{Digestibility (\%)} = \frac{\text{Feed digested}}{\text{Feed eaten}} \times 100$$

$$= \frac{\text{Feed eaten} - \text{faeces excreted}}{\text{Feed eaten}} \times 100$$

That is, the less the amount of faeces excreted for each unit of feed eaten the higher is the digestibility of that feed. *Digestibility is most usefully expressed in terms of D-value, defined as the content (per cent) of digestible organic matter in the dry matter of the feed* (for it is only the organic constituents of a feed that the animal can use for productive purposes).

As well as the energy wasted in the faeces, between 15 and 20 per cent of the energy content of the feed eaten is also lost in the urine that is excreted and in the gases (methane) that are expelled from the rumen during the digestion process. These energy losses are then subtracted from the energy contained in the original feed to give the energy that is available for the animal's maintenance and production processes. This is defined as the Metabolisable Energy (ME) of the feed (Figure 3.1).

Metabolisable energy, which is used as the basis for ruminant rationing systems in the UK, can be calculated from the D-value of a forage with considerable accuracy, using the data for different forages in Table 3.1.

Most measurements of the digestibility of ruminant feeds have been made using sheep, in experiments in which the amount of food eaten and of faeces excreted are both weighed over a balance period, generally of 7–10 days. This is an expensive procedure, however, and is impractical for the farmer or advisor who wants

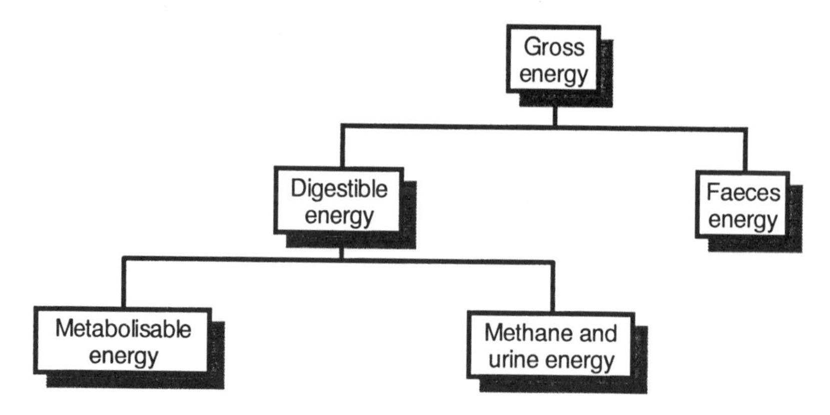

Figure 3.1 Energy losses in the digestive process in ruminant animals

to assess the digestibility of a particular lot of hay or silage. Thus much research has been directed at developing laboratory methods for estimating feed digestibility by laboratory analysis of a small sample of the feed. The most accurate method is *in vitro* digestibility, in which the sample is incubated with liquor taken from the rumen of a sheep, so as to simulate the process of rumen digestion. However, this method is not well suited to routine use, and it has now been largely superseded by the method of Near

Table 3.1 Calculation of the Metabolisable Energy (MJ/kg DM) content of forages from their D-values

Grass hay: field-cured	ME = 0.109 D + 2.63
barn dried	ME = 0.146 D + 0.60
(Moss and Givens (1990); ADAS Feed Evaluation Unit)	
High-temperature dried grass	ME = 0.195 D – 1.82
(Givens, FEU, 1992)	
High-temperature dried lucerne	ME = 0.190 D – 1.40
(Givens, FEU, 1989)	
Grass silage	ME = 0.120 D + 2.91
(Givens, FEU, 1989)	

(Data taken from Alderman, G. (1993) (ed.), *Energy and Protein Requirements of Ruminants*)

Infrared Reflectance (NIR) spectroscopy, described on p. 43. This is a rapid method of analysis which can be carried out on both fresh and dried feed samples; in addition to measuring the D-value and the Metabolisable Energy content of a feed NIR can also measure its moisture and protein contents, as well as the fermentation characteristics of silages.

THE DIGESTIBILITY OF FORAGES AT CUTTING

The equipment for NIR analysis is expensive, and for this reason it is mainly used by central advisory laboratories to assess the nutritive value of conserved forages that are to be fed in the winter feeding programme. This information is, of course, only available after the crop has been conserved; thus an important observation is that a useful estimate of the nutritive value of a conserved forage can be made from a knowledge of the crop from which it was made, the stage of maturity at cutting, and the particular conservation method used.

It has long been known that, as the date of cutting of a forage crop is delayed, it becomes more mature and stemmy, and its digestibility falls. Work at a number of research centres during the 1960s showed that the digestibility of the growing crop could be predicted with considerable accuracy from a knowledge of the forage species and its stage of maturity, particularly during the first main period of growth in the spring. In the case of the grasses, 'maturity' is here defined by the number of days before or after the date on which the particular crop reaches 50 per cent ear-emergence (the stage at which ears have emerged from half the shoots that are going to produce ears). This, of course, applies only to grass; but as grass makes up the bulk of the forage crops that are cut in the UK, the relationships between crop maturity and digestibility are widely applicable.

This research led to the setting up of a systematic programme, during the 1960s and 1970s, to measure the changes in D-value with maturity of all the NIAB-Recommended forage varieties (see Chapter 4). This showed important differences in digestibility between grass species (for example, that cocksfoot was consistently less digestible than ryegrass) and between maturity types within a species (thus 'early' ryegrass was always less digestible than 'late' ryegrass when harvested on the same date).

Unfortunately, with the cut-back in research funding since the early 1980s, this comprehensive study of the yield and digestibility of Recommended varieties was abandoned and, as a result, much less information is now available on the grass varieties that are currently grown. However, the underlying relationships between yield and digestibility are likely to be very similar to that in Figure 3.2, which is based on the widely grown, though now superseded, S.24 ryegrass. This shows that as the crop becomes more mature its D-value falls, initially quite slowly, and then more rapidly after 50 per cent of the ears have emerged. At the same time the yields of both dry matter and digestible organic matter increase. Thus grass harvested at a high yield will be of low digestibility; conversely, when it is cut at a less mature stage it will have a higher D-value but the yield will be lower.

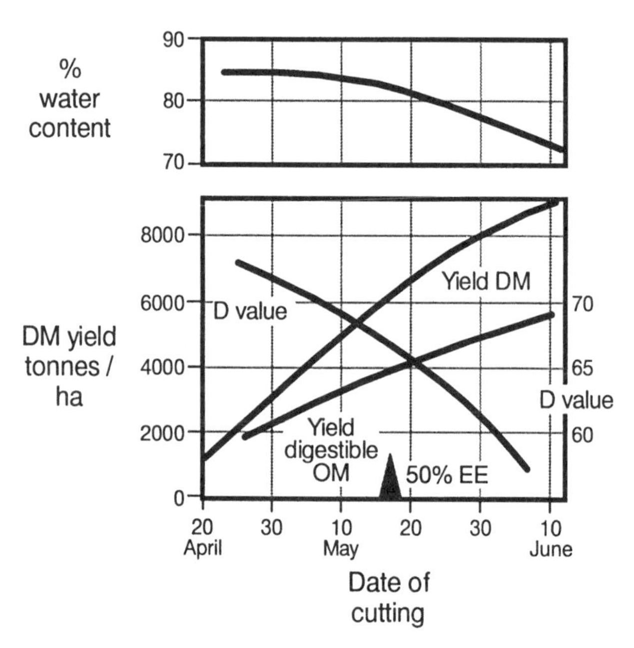

Figure 3.2 Typical changes in yield, digestibility and moisture content of S.24 ryegrass during first growth in the spring (Data: AGRI, Hurley)

The yield and digestibility of the crop when it is cut largely determine the yield and digestibility of the hay or silage that will be made for, as is shown in the following section, the digestibility of well-made hay or silage can be very similar to that of the crop that is conserved. Thus this information can be used to plan when each crop should be cut so as to make hay or silage of the digestibility required for the particular stock to be fed.

Clearly the decision about what crop to grow and when to cut it will also depend on the other feeds that will be available – and on the method of conservation that will be used, for young, high-digestibility grass is in general more difficult to conserve than more mature, but less digestible grass. But the important point is that information such as that in Figure 3.2 is a useful guide to the stage of maturity (date) at which each particular crop should be cut to provide forage of the required D-value. Furthermore, although the information on current grass varieties is less detailed than in the past, the basic principles shown in Figure 3.2 still underpin the decision as to when each crop should be cut to give the optimum balance between yield and digestibility.

A further feature of current forage production is that, compared with the mainly grass-species swards that were sown up until the late 1970s, many more sown swards now include legumes, because of the contribution they make to the nitrogen nutrition of the sward (p. 71). The most commonly sown legume, white clover, has a higher D-value and crude protein content and better intake characteristics than any of the grass species. As a result the nutritive value of grass/clover mixtures is higher than that of pure grass swards fertilised with N – though they may be more difficult to conserve as silage because of their lower content of water-soluble carbohydrate (p. 20).

Figure 3.2 also relates only to the first growth of forage in the spring. While a high proportion of the total annual harvest of hay and silage is made from such first-cut material, much conserved forage is also made from regrowths, either from a previous grazing, or from a hay or silage aftermath. Prediction of the digestibility of these regrowths is less precise than with first growth material, because it depends greatly on when the first cut was taken or when grazing ceased. Thus, again referring to Figure 3.2, up to about 5 May few of the seedheads in a crop of S.24 ryegrass will be more than a few centimetres above ground level, and cutting or grazing before that date will remove few of the potential seedheads. This means that the regrowth from S.24 harvested in early May will contain the full complement of seedheads, so that its digestibility will fall at much the same rate as that of the first-growth material shown in Figure 3.2. In contrast, later in May many of the seedheads will be above cutting or grazing height; thus although the rate of regrowth from cutting or grazing may then be slower than from an earlier harvest, the regrowth will contain few seedheads and its digestibility will fall more slowly.

Long-term leys and permanent swards now make up a higher

proportion of the forage cut for conservation than they did in the 1970s, when a much bigger acreage was sown to leys. This older grassland poses a problem because it is generally of mixed botanical composition. However, just as with sown swards, permanent grass becomes less digestible as it becomes more mature. It also often contains late-flowering strains of grasses and as a result its digestibility is likely to be similar to that of a late-maturity sown grass cut on the same date – though its yield is likely to be lower than that from a vigorously growing young ley.

THE DIGESTIBILITY OF CONSERVED FORAGES

The importance of this information on the digestibility of the crop at the time it is cut is that it can be used to predict the digestibility of the hay or silage that will be made from that crop. Many experiments have shown that, with an efficient conservation method, the digestibility of a conserved forage is very similar to that of the crop from which it was made.

But the proviso 'with an efficient method' is vital; inefficient conservation can lead to a serious fall in digestibility. Thus prolonged wilting in the field, particularly with intermittent rain, can seriously reduce the digestibility of the crop through loss of leaf and the leaching out of soluble constituents; the digestibility of hay falls if it heats up because it has been stored at too high a moisture content; and the digestibility of silage is reduced if it overheats as a result of poor consolidation and sealing, or if large amounts of nutrient-rich effluent are lost from the silo.

Table 3.2 indicates the extent of the fall in digestibility that may be caused by different conservation methods; because the most serious losses are likely to occur during wilting in the field the fall in digestibility is generally greater with hay than with silage. A useful prediction of the digestibility of a conserved forage can thus be made from the estimated D-value of the crop, based on its species and the stage of maturity at which it was cut, together with a correction for the likely fall in digestibility resulting from the particular conservation process used.

This information will be of most use in deciding when each crop should be cut to produce a conserved feed of the digestibility required for the planned winter feeding. However, this can only be an approximate estimate, and the detailed feeding programme will generally require confirmation by laboratory analysis of a sample of the conserved forage, using the relationships between

Table 3.2 Corrections to the estimated D-value of cut forage to allow for losses in the conservation process

Method of conservation	*Subtract from D-value of forage*
Barn-dried hay, good wilting	2
Barn-dried hay, moderate wilting	4
Rapid field hay, good weather	3
Rapid field hay, bad weather	5
Traditional hay, good weather	3
Traditional hay, bad weather	8
Low dry-matter silage, little effluent	0
Low dry-matter silage, some effluent	1
High dry-matter silage, good wilting	2
High dry-matter silage, moderate wilting	3
High-temperature dried grass, direct cut or good wilting	1
High-temperature dried grass, moderate wilting	2

feed composition and digestibility already described. Research on these relationships, much of it carried out at the ADAS Feed Evaluation Unit at Drayton, has shown particularly close agreement between the digestibility of a forage and the measurement of its Near Infrared Reflectance (NIR) – as with silages, shown in Figure 3.3. NIR has thus been widely adopted as a practical

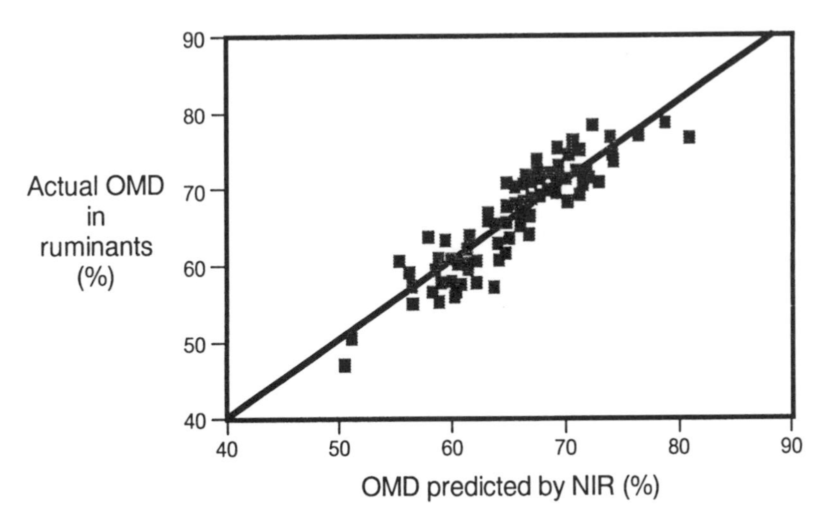

Figure 3.3 The relationship between the digestibility of the organic matter in silage estimated by Near Infrared Reflectance spectrometry and by in vivo *measurement with sheep*

and accurate method of predicting the nutritive value of both fresh and conserved forages, and standard analytical procedures have been agreed between the feed trade and the Advisory Services. The results are generally reported in terms of Metabolisable Energy, which is very closely related to the D-value of the feed (see Table 3.1).

As expected, silage, because it is generally made from crops cut at a less mature, higher-digestibility stage, has a higher D-value and Metabolisable Energy content than hay. However, for reasons not yet fully understood, the ME in silage appears to be less efficiently used than the ME in hay, particularly when both are being fed as sole feed; as a result the ME content of silage may not be a reliable measure of its productive value. Thus Table 3.3 reports the results of a feeding trial in which liveweight gains by beef cattle fed early-cut silage were compared with those by cattle fed late-cut silage, either alone or supplemented with barley. As expected, gains were higher on the early-cut than on the late-cut silage when both were fed as sole feed. However, while the intake of ME on the early-cut silage was slightly higher than that on the late-cut silage supplemented with barley, daily gains were lower; to achieve similar gains the cattle fed the early-cut silage would have had to have eaten 20 per cent more ME above maintenance than they did. Work in Northern Ireland has also shown that the efficiency of utilisation by dairy cows of the ME in diets containing a high proportion of both wilted and unwilted silages is lower than that in similar diets containing dried forages.

Table 3.3 Daily liveweight gains by beef cattle fed on early-cut silage, or on late-cut silage with and without a barley supplement

Silage	*Early cut*	*Late cut*	*Late cut + barley*
Barley (% of diet)	0	0	56
ME intake (MJ/day)	73	60	70
Daily gains			
liveweight (kg)	0.66	0.37	0.80
energy (MJ)	12.2	5.5	14.6

(Data: Thomas, C., Gibbs, B. G., Beever, D. E. and Thurnham, B. R. (1988) (IGER))

The reasons for the lower-than-predicted efficiency of energy use when silage is fed are still unclear. In part, as noted in the following section, it may be the result of a deficiency in protein supply (thus small amounts of protein supplements, such as fishmeal, can markedly increase the efficiency of lean-tissue gain in beef cattle fed on silage); animals may also use more energy in the process of digesting and moving the forage through the gut than when rations high in concentrates are fed. In practice, however, conserved forages are seldom fed as the sole feed to ruminant animals, and the nutrition software used in ration formulation is regularly updated, on the basis of the latest experimental data, to correct for the divergences in efficiency of energy use when conserved forages are fed in mixed rations.

THE PROTEIN VALUE OF CONSERVED FORAGES

Traditionally the protein content of a feed has been expressed in terms of its crude protein (CP) content, calculated by multiplying the total nitrogen (N) content in the feed by 6.25. Forages vary widely in CP content, ranging from less than 6 per cent in mature hay to over 30 per cent in young, leafy grass. Protein content falls steadily as a forage crop becomes more mature; it also varies with the forage species – legumes in general contain more protein than grasses – and with soil N status – N fertilisers increase the protein content of the crop that is grown. These differences are reflected in the CP content of conserved forages, although hay and silage generally contain less CP than the crop from which they were made because of the differential loss of high-N fractions – in particular of leaves and soluble components – which can occur during field treatment and subsequent storage.

However, while efficient processing of a young leafy crop can produce a conserved forage of high protein content, it has long been recognised, particularly in the case of silage, that the productive value of this protein is often lower than would be expected. A major research effort over the last 20 years, which has revealed the complex processes of protein digestion in the rumen shown in Figure 3.4, helps to explain this effect.

The protein in any feed is made up of two main fractions, a soluble fraction which, when it enters the rumen, is broken down into simpler N compounds, and a fraction which passes, largely unchanged, through the rumen and into the small intestine. The former, rumen degradable protein, in turn contains two fractions, quickly degradable protein and slowly degradable protein; these are broken down in the rumen into simpler N-containing com-

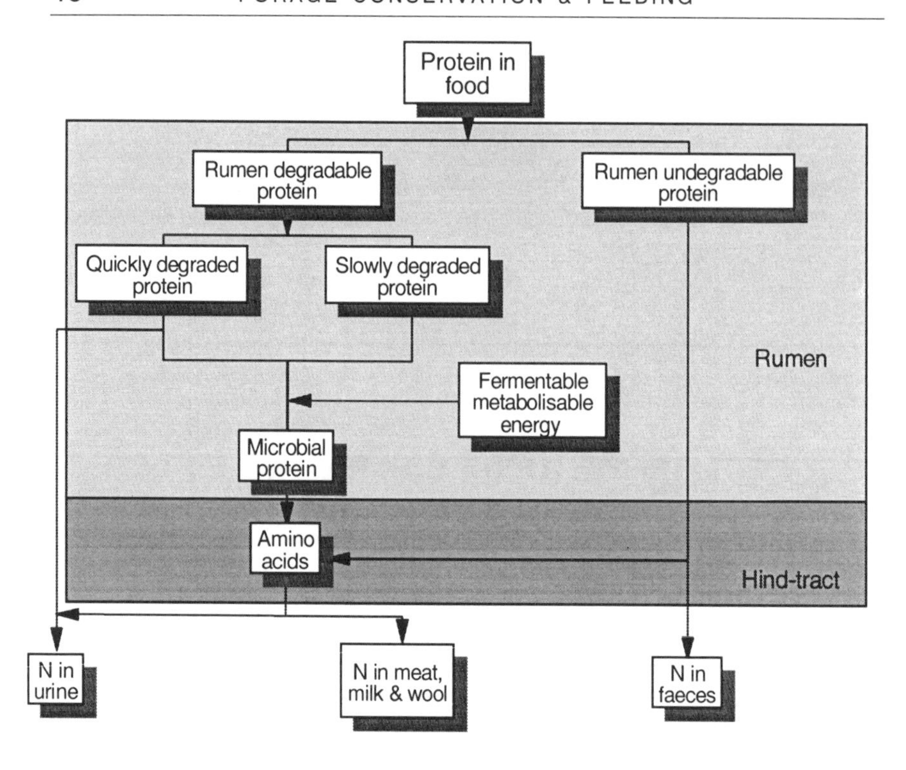

Figure 3.4 The patterns of metabolism of feed protein by ruminant animals

pounds, which the rumen micro-organisms are able to utilise as they multiply in carrying out their essential task of digesting the fibrous part of the feed. The micro-organisms then pass out of the rumen with the undigested food residues, and the microbial protein they contain is then broken down into amino acids in the hind-tract, and absorbed into the bloodstream to provide a major part of the amino acids needed by the animal for growth and milk production. The other part of the protein in the feed, the non-rumen degradable protein, passes through the rumen directly into the hind-tract; there part of it is also digested and absorbed as amino acids, while the non-digested part is excreted in the faeces.

Two important factors influence the amount of microbial protein produced, and so the efficiency with which the rumen degradable protein in the feed is utilised. First, if too much of the feed protein is in the quickly degradable form, it may be broken down into simple N compounds in the rumen faster than the microbes can use them; the surplus N compounds, mainly in the form of ammonia, are then absorbed from the rumen and excreted in the urine as urea, and so are wasted. Second, the microbes must be provided with enough of a suitable energy source (termed fermentable metabolisable energy (FME)) to be able to utilise the degradable protein. The most effective source of FME is the starch in cereal feeds, in particular in ground barley and wheat, which is rapidly fermented by the rumen organisms; the residual starch in maize silage is rather less effective, and some of it passes through the rumen and is digested in the hind-tract.

Much of the crude protein in fresh forage is already in a highly soluble form, and during the process of ensilage some of this protein is broken down still further into simple non-protein-nitrogen compounds, which are immediately available to the rumen micro-organisms when the silage is fed. Thus silage may supply an excess of quickly degraded protein, with the result that the efficiency of use of the crude protein in silage may depend critically on the accompanying energy supply to the rumen. The rumen organisms cannot use the acids in silage as a source of FME (for these acids are the products of a fermentation that has already taken place in the silo); as a result, the protein in silage is likely to be used with low efficiency unless it is supplemented with a good supply of FME – for example the barley supplement shown in Table 3.3.

The efficiency of use of the protein in silage can also be slightly improved if the extent of protein breakdown during the ensilage process is reduced, for example by wilting the crop before it is ensiled, or by using an additive such as formalin (p. 28), both of which reduce protein breakdown. These silages may also retain some of the soluble carbohydrate present in the original crop, as a potential source of FME.

The crude protein in hay is in general less soluble than the protein in silage, so that it tends to be more efficiently used than the protein in silage made from a similar crop. However, the most marked difference is with green crops dried at high temperature (p. 15). Part of the protein in the dried crop is still readily degradable in the rumen, but much of it will have been chemically changed by the heating process, and so is 'protected' from digestion by the rumen organisms. This undegradable protein can thus pass unaltered through the rumen and into the hind-tract, where it is digested and used directly as a source of amino acids. As a result high-temperature dried forages have a higher protein feeding value than either hay or silage.

FEED INTAKE

The other main factor determining the productive value of a feed is the amount of it that livestock will eat. Under many practical situations the contribution that a conserved forage makes to the total ration is limited by the small intake of the forage by the animals being fed. As a result large quantities of supplementary feeds may be needed to achieve a satisfactory level of animal production.

Many theories have been put forward regarding the factors that influence the amount that ruminant animals will consume, given free access to food. First it must be noted that, although some feeds are clearly 'unpalatable' – stock will not eat mouldy hay because they do not like it – this is seldom in practice the main reason for low feed intake. The real problem arises with feeds which animals obviously find palatable, but where other factors appear to limit the amount of feed they are able to eat.

With dry feeds the most important factor is low digestibility; observant stockmen have long known that animals eat much more of young, leafy hay than of old, stemmy hay. The reason is that intake is largely determined by 'gut fill'. When the rumen is full the animal stops eating until some of the food in the rumen has been digested and the residue passes on to the hind-tract, so making space for more food. High-digestibility foods are more rapidly digested, and leave less indigestible residue, than low-digestibility foods. As a result the rumen empties faster, and the animal is thus able to eat more of a high- than of a low-digestibility forage.

Thus high digestibility in feed has two positive effects because, as feed becomes more digestible, both the energy value (ME content) and the voluntary intake of the feed increase. There is a marked response in the level of animal production; as a result early-cut hay has given higher cattle gains, compared with later-cut hay, than would be expected just from its higher digestibility (Table 3.4).

Differences in intake between forages

At one time the relationship between intake and digestibility was thought to be so precise that it should be possible to predict how much of a given forage an animal would eat merely by knowing the digestibility of that forage – in other words, that nutrient

Table 3.4 D-value, intake, estimated ME intake and daily liveweight gains by beef cattle fed on early- and late-cut hay

	Early-cut hay	*Late-cut hay*
D-value	70	60
ME value (MJ/kg DM)	11.2	9.6
Intake (kg DM/day)	3.9	3.4
Intake of ME (MJ/day)	44	33
Liveweight gain (kg/day)	0.9	0.7

(Data: IGER)

intake could be 'read off' directly from a graph such as that shown in Figure 3.2. However, more detailed study showed that there can be considerable differences in voluntary intake between different forages, even though they are of the same digestibility. The most notable example is in the higher intake of legumes (white and red clover, lucerne and sainfoin) than of grasses of the same digestibility. Even among the grasses there are differences which are large enough to be of practical importance – thus the intake of Italian ryegrass is generally higher than that of perennial ryegrass.

The main reason for these differences in intake between forage species appears to be the same as that which relates intake to digestibility, namely that the more rapidly a feed is digested and leaves the rumen the more rapidly is space made available in the rumen for more of that feed to be eaten. Thus it has been shown that animals digest legumes and Italian ryegrass more rapidly than perennial ryegrass of the same digestibility, and so are able to eat more of them.

While the effect of forage species on voluntary feed intake is in general less significant than that of digestibility (though cattle can eat some 40 per cent more dry matter of sainfoin than of timothy of the same D-value!), there is clearly an advantage in growing forages of good intake potential, for example by including (high-intake) legumes in sown mixtures. Forage crop breeders are also now selecting for improved intake in forages, using laboratory screening techniques which closely simulate the rate at which forages are digested in the rumen.

The intake of silage

It was not until the early 1960s that it became evident that the intake of silage depended on factors other than digestibility and rumen fill, which mainly determine the intake of hay. The reason is that most of the research on feed intake had been carried out with 'dry' feeds, and the few comparisons that had been made between hay and silage had been confounded by differences in digestibility, because the hay had generally been made from a more mature, lower D-value crop than the silage. The first relevant comparison seems to have been by Stanley Culpin at Drayton EHF. He found that beef cattle, fed *ad libitum* on either hay or silage *made from the same crop cut on the same day*, gained at 0.68 kg per day on the hay but only 0.57 kg per day on the silage. The reason was that, although the hay and the silage were of almost the same digestibility, the cattle had eaten considerably more dry matter of the hay than of the silage.

The work at Drayton set in train a considerable research programme which showed that the intake of silage can vary over a huge range, with the intake of some silages being similar to that of hay while in other cases silage intake may be 60 per cent lower. In practice there is only a very weak relationship between the intake and the digestibility of silage; in particular the intake of silage of high D-value is often very low.

The research also showed that, while most low-intake silages are also of low dry-matter content it is not high moisture *per se*, but the fermentation products contained in wet silage that are the main cause of low intake. Wilkins and his colleagues at Hurley showed that no single factor could explain all the observed cases of low intake of silage, and suggested that two main groups of factors are involved. First, the intake of high-moisture, well-preserved silage (pH below 4.0) appears to be limited because of the large quantity of acid that the animal has to consume with each unit of silage dry matter it eats. In contrast the intake of wet silage of high pH (often above 5.0) appears to be limited by the metabolic effects on the animal of chemical constituents in the silage, in particular the products of clostridial decomposition of the forage protein (p. 25) – indicated by a high content of ammonia in the silage. In the intermediate range of silages (pH between 4.0 and 5.0), intake is very variable and is not limited by any one chemical factor; thus high ammonia content, though indicating low intake in high pH silage, has shown little correlation with intake when silages are of lower pH. Later research has thus examined combinations of factors. This originally required routine analysis for several chemical constituents in addition to ammonia, which made the procedure time-consuming and expensive. However, a number of centres, including the SAC at Auchincruive and

Hillsborough in Northern Ireland, have recently introduced more rapid methods. These include a technique, based on work in Finland, in which silage juice is titrated with alkali to give a distinctive curve which is closely related to the products of silage fermentation. Wet silage samples can also be directly analysed by NIR (p. 43); this technique can account for some 80 per cent of the variation in intake between different silages, and is now the most widely used for the routine evaluation of silages by both the Advisory Services and commercial companies.

While these new techniques can account for much of the variation in silage intake, they have still not fully identified the specific factors which limit intake. The results have, however, confirmed the key importance, on the one hand, of avoiding very high acid production during the ensilage process, and on the other of minimising protein decomposition, by preventing clostridial fermentation. They thus confirm the priority that must be given either to wilting the crop before it is ensiled, or to using an additive in order to reduce both the extent of protein breakdown and the need for very high acid production during the ensilage process.

Other factors affecting forage intake

The intake of forage is, of course, not solely determined by its digestibility and chemical composition. The amount of silage that small ruminants, including sheep and young cattle, are able to eat can be limited by long chop-length in the silage, and the amount of maize silage that adult cattle can eat also increases at shorter chop-lengths (fortunately short-chopping also improves the efficiency of the ensilage process (p. 131)). Management so as to ensure unrestricted access to feed is also important, including provision of adequate space at feeding troughs and at the silage face with self-feeding, cutting down high or very dense silage faces to make the silage more readily available, and regular removal of stale or soiled feed. The key to success is careful observation of the feeding behaviour of the animals that are being fed.

The effect of the physical form of the feed is most marked with dried forages; in particular there is a big increase in voluntary intake when the particle size in the forage is reduced before it is fed. Experiments with 'dried grass' have given much information on feed intake. Thus dried grass is generally pelleted before it is fed, either directly or after hammer-milling. This reduces the particle size in the feed, which then passes much more rapidly through the rumen than the equivalent unpelleted dried forage; as a result animals eat much more of the pelleted forage than of the 'long' forage.

Table 3.5 shows the resulting effects on the intake and performance of young steers, fed *ad libitum* on either chopped or pelleted dried perennial ryegrass, cut from the same sward at three stages of maturity. As the date of cutting was delayed the liveweight gains by the cattle fed the chopped dried grass decreased, because both the D-value and the intake of the dried feed fell with increasing maturity. At each date of cutting the intake, and the resulting daily gains, were higher on the dried crop that was pelleted before it was fed than on the chopped crop. However, although the increase in intake, and so the benefit from pelleting, was greater with the more mature feeds, liveweight gains on these feeds when pelleted were lower than on the least mature feed fed in the chopped form; the higher intake resulting from the smaller particle size did not compensate for the lower digestibility of the later harvested feeds. A similar improvement in feeding value is found when hay is milled and pelleted; however, the economic benefit is likely to be too small to make this operation profitable.

Table 3.5 Daily liveweight gains by four-month-old steers fed dried grass cut on three dates in spring, and either chopped or pelleted

Date of cutting	*12 May*		*3 June*		*28 June*	
D-value of grass	73		67		59	
	Chop	Pellet	Chop	Pellet	Chop	Pellet
Intake (kg DM/day)	3.72	4.00	3.00	3.72	2.58	3.17
Daily gain (kg/day)	1.00	1.18	0.77	0.95	0.45	0.63

(Data: Tayler and Lonsdale, GRI)

CONSERVED FORAGES FED IN MIXED RATIONS

An understanding of the factors determining the digestibility and voluntary intake of different conserved forages has been an essential stage in the development of effective feeding systems. But only a step; for in practice conserved forage is seldom given as the sole feed, but is fed either mixed with other forages or supplemented with concentrate feeds. It is therefore important to know how these different feeds interact with each other, with the practical aim of achieving the required level of animal production from a ration containing a high proportion of conserved forage, using supplementary feeds to make up the nutritional deficiencies in the forage.

The most common deficiency is in energy intake, and most supplementary feeds contain cereals and other highly digestible carbohydrates to provide extra energy. When these feeds are eaten much of this carbohydrate is rapidly fermented by the rumen micro-organisms to provide the FME (p. 47) that they require; this fermentation also produces organic acids so that the pH in the rumen falls. As the rumen becomes more acid the fibre-digesting organisms become less active and, as a result, both the rate and the extent of fibre digestion fall. Much of this fibre is in the forage component of the ration. Thus cereal supplements, which make the rumen more acid, reduce the rate of digestion of the forage so the animals eat less forage. The extent of the reduction in forage intake for each kg of concentrate fed is termed the 'substitution rate'; this depends on both the 'quality' of the forage and the type of concentrate that is being fed. As Table 3.6 shows, in general the higher the quality (digestibility) of the forage the greater is the substitution rate – concentrates reduce the intake of high-digestibility forages more than the intake of lower-digestibility forages.

Table 3.6 The substitution rate (kg reduction in forage DM intake per kg DM of supplement fed) with different forage types

Forage type	Substitution rate
Poor hay	0.17
Poor grass silage	0.32
Lucerne hay	0.44
Average grazing	0.55
Dried grass	0.55
Zero grazing	0.6–0.7
Medium grass hay	0.63
Maize silage	0.63
Good grass silage	0.68
Dried lucerne 'wafers'	0.78

(Data: Bines, NIRD)

These substitution effects are less clear-cut when silages are fed because, as has been noted, the intake of silage is determined more by the fermentation characteristics in the silage rather than by its level of digestibility. However, silages with the highest intake potential when fed as sole feed (irrespective of their level of digestibility) tend to show the highest substitution rates when concentrates are fed with them (Table 3.6). In contrast, the (low) intake of poorly fermented silage, even if the silage is of high digestibility, is not greatly reduced when concentrates are fed.

The partial replacement of forage when a supplementary feed is given inevitably means that the resulting response in animal performance is in general less than would be predicted from the additional nutrients supplied by the supplement. In particular the benefits of the higher digestible energy content in early-cut forage may not be fully realised when concentrates are also being fed. Thus if 1.0 kg of concentrate fed to a dairy cow reduces its silage intake by 0.5. kg, the resulting response in milk yield, at 0.8–1.0 kg per day, is considerably less than the theoretical response of 2.0–2.5 kg. It also appears that the substitution rate may be higher at higher levels of concentrate feeding; thus with dairy cows in early lactation a substitution rate of 0.2 kg of silage dry matter per kg of concentrate was found when a low level of concentrates was fed, but this increased to 0.9 kg at high-concentrate feeding (in other words, the intake of silage DM fell by 0.9 kg for each extra 1.0 kg of concentrate). As a result the benefits of early cutting on total energy intake and on animal production may be more evident at low than at high levels of concentrate feeding (see also Figures 10.2 and 10.5).

In many cases concentrate supplements are also fed to make up a deficiency of metabolisable protein in the basic forage diet. This deficiency can be quantitative – the forage just does not contain enough protein for the animals' requirements – but increasingly it is the *quality* of the forage protein that is deficient, and that the concentrate must compensate for. As already noted in Figure 3.4, even when the forage appears to contain enough crude protein, much of this may be wasted because the protein is broken down in the rumen into simple N compounds more rapidly than these can be used by the rumen organisms. In that case utilisation of N can be improved by giving supplementary feeds containing sugars and starch, which increase the supply of FME in the rumen.

However, the quantity of amino acids absorbed from the hind-tract may still not be enough for the animal's metabolic requirements; in that case the most effective supplements are those with a high content of undegradable protein, which passes unchanged into the hind-tract. In a feeding trial at the GRI a supplement of 150 g of fishmeal increased the daily body-protein gain of cattle fed on silage from 95 kg to 145 g. Dried grass is also an effective supplement to silage because part of the protein it contains has been protected from rumen digestion by the heating process

(p. 47). Such 'rumen undegradable protein' is now an important constituent of many concentrate feeds, in particular for high-producing animals. Rations for very high-producing animals are also being supplemented with specific essential amino acids, including methionine and lysine, which have been treated so that they are 'protected' from digestion in the rumen but can be digested and absorbed in the hind-tract.

Hay and silage made from very mature crops are likely to be deficient in both energy and protein, and the type of protein supplement fed is then less critical. In contrast, while silages made from forage maize (p. 76) and from whole-crop cereals (p. 81), provide a useful source of FME, this energy may be inefficiently utilised because these silages are of low crude protein content and do not supply enough soluble N compounds for the requirements of the rumen micro-organisms. This deficiency of N can be at least partly compensated by feeding supplements containing a readily available source of non-protein N, such as urea, or by mixing urea or ammonia with the crop during the ensilage process (p. 82). In the rumen the urea is rapidly broken down to ammonia, which the micro-organisms, provided with a good supply of FME from the cereal silage, can convert to microbial protein, which then passes on to be absorbed as amino acids in the hind-tract.

The composition of the diet determines the pattern of rumen fermentation, and this in turn can also affect the composition of the animal product (p. 203). Thus diets high in fibre, such as forages, promote the production of acetic acid in the rumen; this acid is then absorbed from the hind-tract and supplies one of the main building blocks in milk-fat synthesis – hence the observed increase in milk-fat percentage when additional hay or straw is fed to dairy cows. Conversely, cereal feeding reduces acetic acid production in the rumen, so reducing milk-fat percentage, but increases propionic acid production, with benefit to milk protein content.

In practical terms the reduction in forage intake when cereal-based supplements are fed is perhaps not important when only limited amounts of hay or silage are available; it does, however, limit the extent to which forages of high-digestibility and high-intake potential can be exploited under *ad libitum* feeding conditions. Increased attention is therefore being given to ways of reducing the effects of concentrate feeding on forage intake. As has been noted, this is mainly due to the reduced fibre-digesting ability of the rumen micro-organisms under the conditions of high acidity (low pH) produced in the rumen when cereal supplements are fed. This effect is probably exaggerated by the common practice of feeding concentrates once or twice a day (as with parlour feeding of dairy cows), leading to rapid production of acid in the rumen. There is evidence that rumen pH is maintained at a higher and less fluctuating level if the cereal supplement is fed more frequently, either from out-of-parlour feeders, or mixed with forage in a complete diet, and that animals fed in this way can eat more forage. An alternative approach is to feed an alkaline supplement, with the aim of partly neutralising the acids produced in the rumen when cereals are fed. Work at Boxworth EHF showed an increase in hay intake, and a response in milk production and in milk-fat percentage, when dairy cows were supplemented with 200 g daily of sodium bicarbonate (Table 3.7); there could be similar benefits from feeding urea-treated wholecrop cereals and alkali-treated straw (p. 200).

Table 3.7 Effect of including sodium bicarbonate in the concentrate ration of dairy cows fed on hay *ad libitum*

	With concentrate	*With concentrate + 200 g bicarbonate/day*
Hay intake (kg DM/day)	7.9	8.8
Concentrate (kg DM/day)	9.0	9.2
Milk yield (kg/day)	22.9	24.3
Milk fat (%)	3.25	3.60

(Data: ADAS, Boxworth EHF)

Fibre in ruminant feeding

Ruminant animals are particularly well adapted to utilising fibrous feeds; but do they need fibre in their diet? Here it is necessary to distinguish between fibre as a chemical fraction in the feed and fibre as the structural component of 'fibrousness'. As a forage crop grows more mature it becomes physically more fibrous; in this process the chemical molecules of cellulose and hemicellulose become linked together by lignin molecules and as a result become increasingly resistant to the fibre-digesting microbes in the rumen – hence the decrease in digestibility with maturity. Yet a certain amount of physical fibre is needed to stimulate cudding, which is essential for effective rumen function, in particular by promoting the secretion of saliva which prevents the rumen contents becoming too acid. Thus acidosis and bloat, which proved a hazard in the early days of 'barley-beef' feeding, were avoided by feeding about 0.5 kg of straw daily with the rolled barley ration (which contained virtually no 'fibre'), so as to ensure that the cattle ruminated properly.

In the case of the dairy cow, by preventing the rumen becoming too acid the fibre in the diet also creates the conditions which stimulate the production of acetic acid by the rumen organisms, so minimising the problem of low butterfat levels in milk when high levels of energy concentrates are fed. Particularly in early lactation, the feeding regime should thus aim to prevent a low pH in the rumen by providing *ad libitum* access to high D-value forages which provide physical fibre, but in which this fibre is not so highly lignified that it is not well digested by the rumen organisms. As nutrient requirements fall later in the lactation more forage can be fed, and there is then less risk of a shortage of 'fibre' in the diet; in practice, the problem then is more often that the forage contains too much fibre, and so is of low digestibility and intake.

The mineral content of conserved forages

The requirements for minerals by different types of livestock, and the content of minerals in a wide range of feedstuffs, are given in standard feeding tables. These tables indicate that, of the major mineral elements, phosphorus, magnesium and sodium, and of the minor elements, copper, cobalt and selenium are the most likely to be deficient in rations containing a high proportion of home-grown forages, and when only small quantities of purchased concentrates, which generally contain mineral supplements, are being fed.

In the case of forages some general trends can be noted – for example, the higher total mineral content of legumes than of grasses; the higher calcium to phosphorus ratio in the legumes; the low content of both sodium and magnesium in timothy. Mineral content also varies considerably in the same forage species grown on different soils, and also with the amount and type of fertiliser applied. Thus potassium fertilisers can reduce the contents of both sodium and magnesium in forage. Such information can indicate when supplementary minerals may be needed. However, feeding tables only give average analyses and, as animal feeding becomes more precise, data on the mineral content of the specific feeds in the ration are needed. While it is impractical for every lot of forage to be analysed for mineral content, there is great advantage in the livestock farmer having mineral analyses carried out on a range of the crops he grows, so as to identify which minerals are most likely to be deficient in the forages he harvests – and to provide the data on mineral content to be incorporated in the input software to his feeding 'programmes'.

Particularly with the wider use of computer-generated feeding programmes, a detailed knowledge of the factors determining the feeding value of the conserved forages and other components of livestock rations might not seem to be necessary. In our view, however, an understanding of the general principles of forage digestion and metabolism, and of the interactions between forages and other feeds, is basic to planning the forage programme. While improved methods of forage analysis now provide, at a relatively low cost, a much better description of feeding value, and this information is increasingly being incorporated in computer nutritional software, the computer can only indicate the best use of the hay or the silage that has been made once it is in the barn or the silo. But the quality of these feeds will depend on the decisions that were made months before – on what crops were grown, when they were harvested, and how they were conserved. In our view these management factors, which make up much of the content of this chapter, will continue to play a key role in profitable livestock farming.

CHAPTER 4

CROPS FOR CONSERVATION

Most of the hay and silage made in the United Kingdom is harvested from fields which are also grazed at some time of the year, so that the crop species that are grown need to be suitable for both cutting and grazing. In many cases grazing is the main method of utilisation, with grass being cut only when it is surplus to the current requirements of the grazing animals, in order to keep the pasture in good condition for future grazing. In other cases most of the forage production from the sward is cut, with only limited grazing, generally later in the season. A few crops are grown for cutting only, including forage maize and wholecrop cereals for silage, and lucerne for high-temperature drying.

Basically, most grass fields, whether they are cut or grazed, have a pattern of herbage growth during the year similar to that shown in Figure 4.1, with the maximum rate of growth during the 'spring flush' in May and June, a check in mid-summer, and then a lift in production during the early autumn. The actual rate of grass growth at each particular time of year depends, of course, on many factors, including the soil nutrient status and fertiliser input, the moisture supply and soil temperature, the botanical composition of the sward, and the timing and frequency of harvesting. However, the general overall pattern of growth is as shown, and this seldom matches the relatively level feed requirement of most livestock enterprises. Thus while the grazing requirements of the spring-calving dairy herd and of fattening cattle and sheep sold off during the summer *do* fall off after mid-summer, in most cases much more forage is grown in the spring than can possibly be utilised by grazing. This 'surplus', often together with a smaller 'surplus' available in the autumn, provides the main supply of forage to be conserved for feeding in the winter, when little grass is available for grazing.

Conserving this surplus grass, generally as hay, has long been an integral part of traditional livestock farming. But high losses from the conservation methods that were used, and the poor

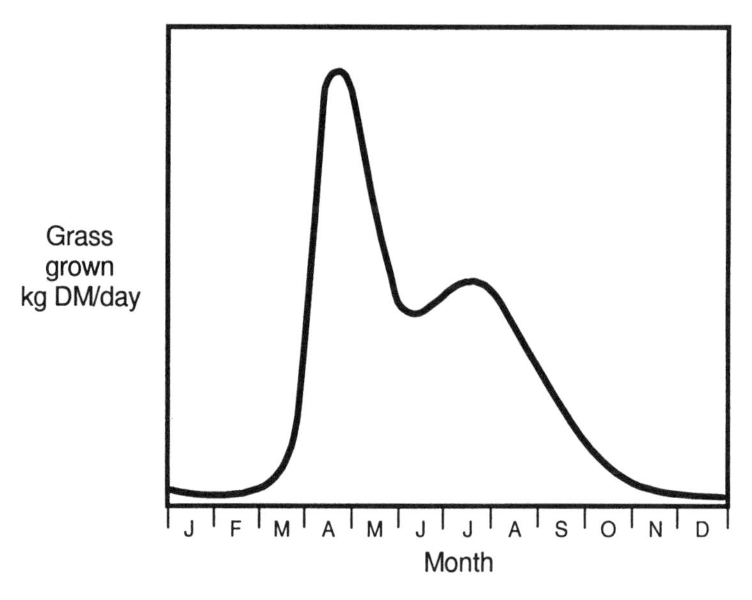

Figure 4.1 Daily dry-matter production through the year from a grass sward fertilised with N and irrigated

feeding value of much of the hay that was produced, made this operation very much of subsidiary importance to grazing as the main way of utilising the grass that was grown.

This inefficiency in forage conservation was seldom challenged – to the extent that in the 1950s a major research programme was set up to develop systems of 'all-the-year-round' grazing, aimed at minimising, or even eliminating, the need for forage to be conserved for winter feeding. Fortunately, and at much the same time, research was also started into better methods of forage conservation, and this led to the development of practical conservation systems which can produce high-value feeds with low losses – and which are the subject-matter of this book. These methods have been widely adopted, and as a result forage conservation now plays a much more important role, both in providing winter feed and in improving the quality of summer grazing, than it did in the immediate post-war period.

Clearly, when the primary purpose of a sward is for grazing, the quality of the herbage that is cut from that sward may at times have to take second place to the needs of the grazing animals. However, in many regimes of grazing management, including most rotational grazing, and more formally in a system such as that shown in Figure 4.2, cutting is fully integrated with grazing so as to ensure an adequate supply of forage, of high quality, for both grazing and cutting. Even under 'continuous' grazing, which has replaced more complicated grazing systems on many livestock farms, a part of the grazing area may be fenced off and cut whenever grass growth is getting ahead of the stock, both to provide winter forage and to maintain the quality of the grass for subsequent grazing.

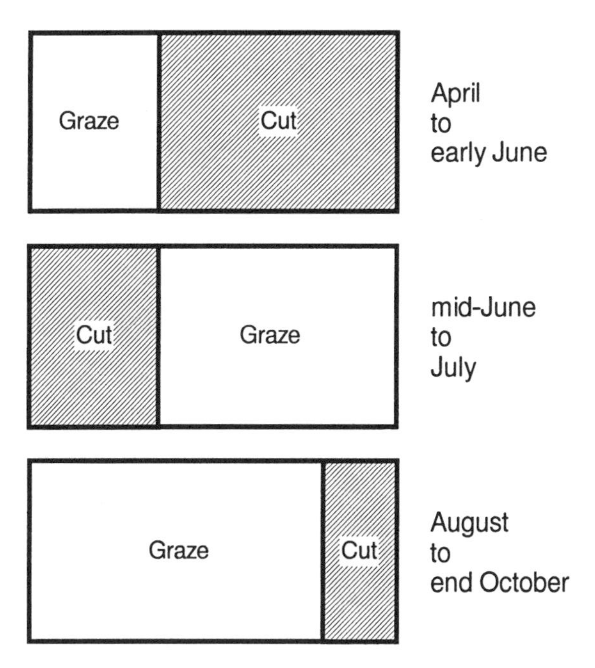

Figure 4.2 The integration of cutting and grazing to give a succession of forage for grazing

Sown grassland (leys) is particularly well adapted to a combination of cutting and grazing, and the information on yields and nutrititive value, outlined in Chapter 3, can assist greatly in its management, and in deciding when each field should be cut to provide forage for a particular feeding purpose. Information, similar to that for S.24 ryegrass shown in Figure 3.1, on the yield and digestibility of all the herbage varieties in the NIAB Recommended National List was readily available up until the early 1980s, both from the (then) Grassland Research Institute and from the National Institute of Agricultural Botany. However, because of the large numbers of 'named' varieties which were being listed, it was then decided to publish most of the data in terms of maturity types *within* each species, rather than for individual varieties. Thus Figure 4.3, based on data from NIAB Technical Leaflet No. 2, published in 1985, shows the changes in D-value, with increasing maturity during first growth in the spring, of the different maturity types of the main grass species, and Figure 4.4 shows the same data for different legume species. These two figures demonstrate a number of important points:

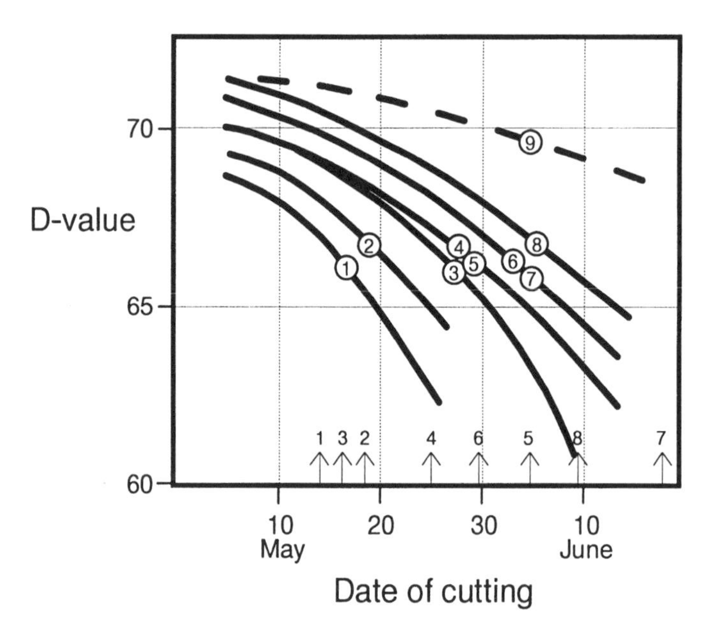

Figure 4.3 Changes in digestibility (D-value) during the first growth of different maturity types of several grass species, and of white clover. Key: 1 cocksfoot, early; 2 cocksfoot, late; 3 perennial ryegrass, early; 4 Italian ryegrass; 5 timothy, early; 6 perennial ryegrass, intermediate; 7 timothy, late; 8 perennial ryegrass, late; 9 white clover. Arrows indicate approximate date of 50 per cent ear emergence (Data: NIAB)

- The D-value of all the grasses decreases as they become more mature, but with an indication that the rate of fall-off in digestibility is rather slower with the later-maturing types.
- On a given date different grass species, and different maturity types within a species, can differ considerably in digestibility. Thus both early and late varieties of cocksfoot are less

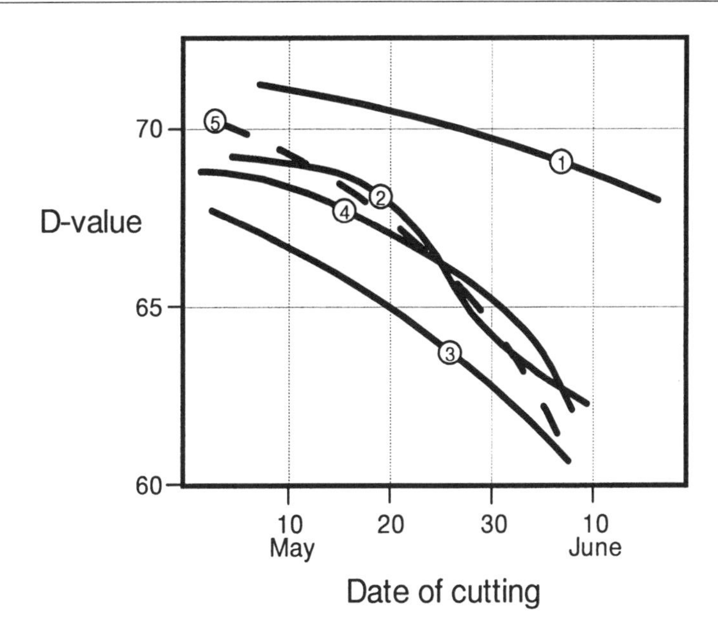

Figure 4.4 Changes in digestibility (D-value) during the first growth of several legume species. Key: 1 white clover; 2 red clover; 3 sainfoin; 4 lucerne; (5 early ryegrass) (Data: NIAB)

digestible than the corresponding ryegrasses, while late-maturing varieties of perennial ryegrass, both diploid and tetraploid, are more digestible than earlier-maturing varieties.

- White clover is always more digestible than any of the grass species; as a result, mixed swards containing white clover have a higher D-value than pure grass swards. The digestibility of red clover and sainfoin is similar to that of the intermediate-maturity perennial ryegrasses, but lucerne is always less digestible than ryegrass cut on the same date.

In making decisions on which forage species and varieties to grow and when to cut them, yield, as well as quality (D-value and protein content), is an important consideration, and an effective way of comparing the yielding ability of the different forage species and maturity types was adopted. Thus Figure 4.5 shows the yield of each of the different maturity types within each grass species on the date on which its digestibility falls to 67 D-value during first growth in the spring. These dates relate to the 'average' season in the south of the United Kingdom; the actual

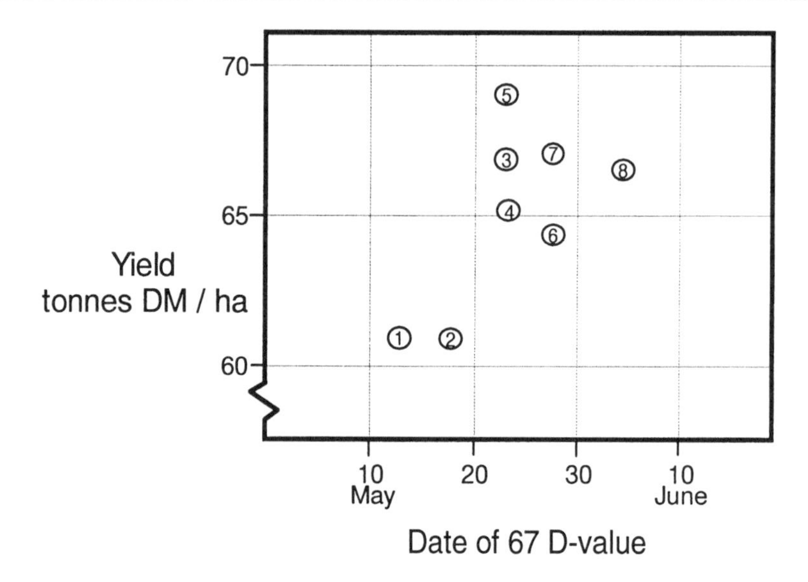

Figure 4.5 Dates at which the digestibility of the first growth of different grasses falls to 67 D-value, and the approximate dry-matter yields on those dates. Key numbers as in Figure 4.3 (Data: NIAB)

date will be later in a 'late' season, and also in the north of the country. The farming press publishes average D-values each spring for different areas, from which corrections can be made for the earliness or lateness of the particular season. 67 D has been adopted as a 'bench-mark' compromise between yield and feeding value, and Figure 4.5 is not intended to indicate that this is the optimum date of harvest/D-value, which will, of course, also depend on many other factors, including the class of stock that are to be fed.

Again a number of important points are seen:

- The digestibility of all the ryegrasses, both Italian and perennial, does not fall to 67 D until towards the end of May, by which time conditions for cutting and wilting in the field are likely to be better than in mid-May, and the digestibility of cocksfoot has already fallen to 67 D.
- Figure 4.5 shows that there can be a 'spread' of more than 12 days between the dates on which the D-value of early-maturing and of late-maturing ryegrass falls to 67.
- As a result of its generally lower digestibility, the yield of cocksfoot at 67 D-value is always lower than that of the other grass species.

- Perhaps unexpectedly, both early and late varieties of timothy give higher yields of dry matter at 67 D-value than any of the ryegrasses. Yet, despite these high yields, and despite timothy being highly palatable to livestock, this species makes up only 6 per cent of the grass seed that is sown in the UK, possibly because it is more difficult to establish than ryegrass.

As noted above, these results have been presented in terms of maturity types (early, medium and late) within each species, at least in part because the increasing number of named forage varieties appearing in the Recommended Lists made publication of data on individual varieties impractical. However, it was also done because the NIAB had found only small differences in either yield or digestibility between the different named varieties within any one species/maturity type that were then being marketed. We commented at the time that this seemed to indicate that relatively little progress had been made in grass breeding, compared with that, for example, in cereal breeding. (In 1986, S.24 perennial ryegrass, which had been released for general use in the mid 1930s, was still the standard against which other ryegrass varieties were being evaluated – equivalent to Little Joss being accepted as the standard wheat variety!)

As a result of changes in the organisation and funding of NIAB, comprehensive measurements, such as those on which Figures 4.3, 4.4 and 4.5 were based, are no longer being made, and access to much of the data that is recorded is restricted. However, the limited data included in the NIAB publication, *Recommended List of Grasses and Herbage Legumes, 1994/95*, do suggest that useful progress has been made, during the last ten years, in breeding for improved yield and quality in both grasses and legumes. In particular:

- The yields of a number of the main grass maturity types on the List, under the standard cutting regimes used in the NIAB trials, are now higher than those that were recorded in 1985.
- The apparent link between 'heading date' and the date at which D-value falls to 67 is now less clear-cut than that implied in Figure 4.3. Thus, within the current varieties of early-maturing perennial ryegrass, Barkate is recorded as 'heading' some 18 days before Sambin, but the digestibility of the two varieties falls to 67 D on the same date, and their yields are then very similar.
- There has also been some success in breeding cocksfoot varieties of higher digestibility; thus the D-value of four of the

varieties in the latest List does not fall to 67 until the third week in May – a week later than the cocksfoot varieties shown in Figure 4.5, and similar to the early ryegrasses.

- Yields of red clover and lucerne are no higher than in 1986, but good progress has been made in breeding both medium- and large-leaved white clovers. These are more persistent than previous varieties when sown in grass/clover mixtures and so can make a bigger contribution to both the N economy and the nutritive value of the sward.
- The 'spread' of dates over which the D-value of varieties of perennial ryegrass falls to 67 has been extended, from 12 days to some 18 days.

This spread may seem small in relation to the full span of spring and summer; but it could be of real practical value in the management of grass fields for both cutting and grazing. If different fields on the farm are sown to early-, medium- and late-maturity types (each with a mixture of varieties and/or species of one maturity type), they should provide a succession of first-growth forages of similar quality for harvesting over a period of two weeks or more, with the crop on the latest-maturing fields still being above 67 D-value in early June.

Sward management, yield and quality

The preceding comments have dealt mainly with the first growth of forage in the spring, because this illustrates most clearly the differences between the different forage species that may be grown. This first harvest is also of critical importance because it generally contributes between 35 per cent and 55 per cent of the total annual production of dry matter from the sward; it provides more than half the total forage that is cut for conservation; and the timing of the first harvest has a big effect on the subsequent regrowth of the sward.

However, it must again be emphasised that the comparison of different forages at 67 D-value (see Figure 4.5) is not meant to imply that 67 D is the optimum level of digestibility to aim for. The main usefulness of this information is in helping the farmer to decide what crops to grow and when to harvest each crop for his own particular livestock enterprise. In doing so he must take account of many factors, which are examined in more detail in Chapter 10, including:

- The class of stock to be fed in the winter; thus dairy and beef cattle will need more digestible feed than suckler cows and store cattle.

- The likelihood that during the winter an autumn-calving dairy herd will require conserved forage of higher digestibility than a spring-calving herd.
- The cost and availability of alternative feeds, including cereals and cereal-based concentrates, and of novel feeds such as alkali-treated straw (p. 167).
- The stocking rate on the farm; at high stocking rates only a small area may be available for cutting, and the best policy may then be to take heavy cuts of relatively low-digestibility forage, and to buy in supplements for winter feeding.
- The rate at which the cut sward will regrow for subsequent grazing or cutting.

The typical changes in yield and digestibility as a 'grassland' crop matures during its first growth in the spring were illustrated in Figure 3.2 (though 'field' yields are in practice likely to be lower than the 'experimental' yields shown in that figure). The most important point to note is that the yields of both dry matter and digestible dry matter increase almost linearly into early June. Thus the yield of a crop cut early to get forage of high digestibility will be lower than that of the same crop, cut later, and so of lower digestibility.

The first cutting (or grazing) also has a major effect on the subsequent rate of regrowth of the sward because it removes much of the actively photosynthesising leaf area. However, the extent of the reduction in the rate of regrowth is much more marked after a heavy (late) cut has been taken than after an earlier cut – at least in part because much more leaf remains, ready to resume growth, after an early than after a late cut.

Soil moisture conditions are also generally more favourable for regrowth after a cut in mid-May than from a later cut in early June. This slower rate of regrowth following later cutting could be critical if the regrowth is needed for grazing stock.

As a result, although late first cutting does give higher yields of both dry matter and digestible dry matter than an earlier cut (see Figure 3.2), the total annual yield is not much greater than when an earlier first cut is taken, *while the average quality (D-value) of the forage may be much lower* (Table 4.1). Harvesting the first cut at 'medium quality' also reduces both total annual yield and overall quality, but on some farms this will be justified by the security of having a large quantity of winter feed in the barn by early June.

Table 4.1 Total annual yield (tonnes dry matter per hectare) and mean digestibility (D-value) of early perennial ryegrass first cut on three different dates in the spring

		Yield	
Cutting date	*High quality (68 D)*	*Medium quality (65 D)*	*Low quality (61 D)*
First growth			
18 May	4.3		
29 May		6.4	
10 June			8.5
Regrowth			
22 June	3.4		
13 July		2.6	
27 July	2.6		
12 August			4.0
Autumn grazing	2.9	3.2	1.8
Total yield	13.2	12.2	14.3

(Data: *Milk from Grass*; AGRI/MMB)

However, on most high-yielding dairy farms the current target is for silage of 65 to 70 D-value, and to achieve this the first harvest from early perennial ryegrass and Italian ryegrass swards has to be taken in mid-May. It is here that the spread of maturity types, shown in Figure 4.3, could be useful, for if part of the area to be harvested is sown to later-maturing ryegrass varieties, that area need not be harvested until early June, when conditions for field-wilting are likely to be better than in May – so spreading both the weather risk and the pressure on men and machines. A later harvesting date, though with some sacrifice in yield, can also be achieved by grazing part of the sward in late April/early May so as to remove some of the potential flowering stems (p. 41). Such early grazing has the further advantage that it helps to maintain tiller density and improves sward persistence and resistance to invasion by weeds.

Another important development has come with the recognition that dairy cows in the second half of their lactation can be much more tightly grazed in the spring, without reducing their milk output, than was thought possible in the past. Such tighter grazing can make a bigger area of first-growth forage available for cutting (see Figure 4.2); either more forage can then be conserved, or the forage can be cut earlier, and so will be of higher D-value. This earlier cutting will also give more rapid regrowth for subsequent cutting or grazing, particularly since soil moisture is likely to be higher than it would be after a later cut, and there should be a faster response to fertiliser N.

Fertiliser use on grass and forage crops

The results of research showing the higher digestibility of the ryegrasses than of (most) other grass species were first published in 1960 (for many farmers this merely confirmed their own experience that ryegrass swards produced better livestock feed than other grass species!). Current varieties of Italian and perennial ryegrass have the further advantage that they continue to yield more than other grass species (Table 4.2). As a result more than 90 per cent of the grass seed now sown in the UK is ryegrass; of this more than 50 per cent is of intermediate and late-heading ryegrass varieties which, although they start active growth in the spring a little later than early varieties, remain above 67 D until the end of May/early June.

Table 4.2 Annual yield of grass species under NIAB 'conservation cutting' regime. Average of 1987/88 data

Species	*Variety*	*Yield (tonnes dry matter per hectare)*
Perennial ryegrass	Talbot	13.8
Italian ryegrass	RvP	15.9
Timothy	Motim	13.0
Cocksfoot	Sylvan	13.4

(Data: NIAB)

A further major change in grassland production over the last 30 years has been the greatly increased use of nitrogen fertilisers, with the average application of N to grassland in England and Wales increasing from about 50 kg N per hectare in 1960 to 135 kg per hectare in 1986. However, since the late 1980s the level of N use has been decreasing, to just over 100 kg N per hectare by 1993. There have been a number of reasons for this:

- The rate of N use per hectare on dairy farms is in general more than double that on beef and sheep farms (Table 4.3). However,

Table 4.3 Average use of fertiliser nitrogen (kg N/hectare) on different grassland farms, England and Wales, 1993

Farm type	*Grass under five years old*	*Grass over five years old*
Dairying	208	163
Beef cattle and sheep	104	71

(Data: *Fertiliser Use*)

following the imposition of milk quotas in 1984 there has been a steady fall in the number of dairy cows in the UK; thus, even though over the same period there has been an increase in the numbers of beef cattle and sheep, overall there has been a reduction in average N use on grassland.

- There has also been a steady fall in the area of sown swards less than five years old, and an increase in the area of older swards (Table 4.4). Less N is applied per hectare on older swards, again leading to a reduction in total N use.

Table 4.4 Areas of grassland of different age ('000 hectares)

Age of grassland	*1981–3 (average)*	*1992*
Under five years	1872	1558
Over five years	5102	5225
Total	6974	6783

(Data: MAFF/HMSO, 1993)

- After several decades during which reliance on N fertiliser meant that few legumes were included in grassland seed mixtures, there has been renewed interest in the potential of legumes, both as a source of N fertility, and in improving the nutritive value of the forage that is grown. This has been encouraged by the success of plant breeders in producing a number of new varieties of white clover which are more persistent and productive than the earlier wild white types, and which can 'fix' more than 200 kg of N per hectare under a dairy cow grazing/silage cutting regime. In recent field-scale experiments (Table 4.5) the yield from grass/white clover swards, cut for silage three times a year (a regime under which the traditional wild white clover would have largely disappeared), was between 82 and 97 per cent of the yield from a grass sward fertilised with 360 kg N per hectare. Most importantly, in the last three years the yield from the clover sward at the critical first harvest was at least equal to that from the heavily fertilised sward.

- Less N is also being applied because of the more detailed practical advice that is now available on the efficient use of N fertiliser. This advice takes account of the soil, aspect and rainfall of the farm (good, medium and poor growing conditions); the N available in the soil, as a result of the previous management; the impact of the current management, with cutting removing much more N from the system than grazing; the more efficient use of farmyard manures and slurry as a result of improvements in storage and application systems; and, most importantly, the imperative of reducing the amount of nitrate leaching into rivers and groundwaters as a result of excessive use of N.

Table 4.5 Annual herbage yields (tonnes dry matter per hectare) from three silage cuts taken from grass swards fertilised with nitrogen, and from grass/white clover swards (1988–91)

Sward	*1988*	*1989*	*1990*	*1991*
Grass +360 kg N/ha	12.0	11.5	11.9	10.9
Grass/white clover	9.9	9.3	11.4	8.9

(Data: Scottish Agricultural College)

On average, the yield of grass increases by about 20 kg of dry matter for each kg of N applied. However, the response is lower under 'poor' growing conditions (for example, on shallow soils in areas with less than 300 mm of rainfall); when the N is applied very early in the season, as under the T-sum regime, in which the date of application is based on the cumulative soil temperature; and at levels of application much above 250 kg N per hectare, because of the effect of 'diminishing' returns.

Unfortunately, despite considerable research, there is still no reliable method of analysing a sample of soil for its 'available N' content. Thus while the principles of optimum N use are now well established, specific advice continues to be based on knowledge of each farm in terms of its particular growing conditions, type of livestock enterprise, and the intensity of production that is aimed for. In most cases the advice is likely to be that N fertiliser should be applied at less than the maximum economic rate (the rate at which the cost of the N equals the value of the extra utilised grass that is grown), because of the difficulty of judging this point under practical field conditions.

It will also be important to prevent environmental damage by observing the MAFF Code of Good Agricultural Practice for Water, Soil and Air; this will almost certainly lead to further reduction in N fertiliser use, particularly in the nitrate sensitive areas of the country, and to more careful use of the N in animal manures. Particular care is needed when N fertiliser is applied to stimulate grass growth early in the spring, with the aim of getting winter-housed stock out to pasture as early as possible, because of the risk of some of the N being washed out of the soil if the application is followed by a cold, wet spell of weather, so that it is not taken up by the growing crop.

In contrast to N content, soil can be accurately analysed for its contents of available phosphorus (P) and potassium (K), expressed either as mg per litre of soil or as an index in the range 0 to 3. Yield responses by grass to P and K fertiliser are only likely at the low indices 0 and 1, and most soils now analyse at a higher index. Thus advisory tables are based on replacing the minerals removed in the harvested grass so as to maintain soil reserves, taking full account of the input from animal manures, rather than on routine applications.

Soil analysis also measures the amount of available magnesium (Mg). This is an important element because, although grass itself seldom suffers from Mg deficiency, the forage that is grown may not always contain enough Mg to prevent hypomagnesaemia (grass staggers) in the livestock that are to be fed. Thus applications of Mg, as magnesian limestone or calcined magnesite, are advised on soils analysing at Mg indices 0 and 1; the quantities applied should at least replace that removed in the crop (5 tonnes of dry matter cut for silage contain up to 20 kg of MgO). This is particularly important where high inputs of N and K are also being applied, because K can reduce the uptake of Mg by the growing crop, while both N and K can make the Mg in the herbage less nutritionally available to the livestock that eat it.

There is also increasing evidence of a deficiency of sulphur (S) in many areas of the United Kingdom, both because control of industrial pollution is reducing the amount of S deposited from the air, and because smaller amounts of S-containing fertilisers, such as ammonium sulphate and single superphosphate, are now used. S deficiency is most likely to occur when annual atmospheric deposition is less than 30 kg S per hectare, particularly on sandy and chalk soils, and under these conditions there is a considerable response to S-containing fertilisers. Most soils provide enough S for the first harvest in the spring, but herbage production has increased by up to 30 per cent in the second and later harvests following application of S on deficient soils. Again soil analysis is unreliable for assessing S deficiency, and analysis of the S content and of the N:S ratio in herbage is generally recommended; a response in herbage production is likely when the ratio of N to S in the herbage dry matter is greater than 13, indicating that the plant does not contain enough S for the synthesis of the essential amino acids cysteine and methionine. Animal manures contain useful amounts of S, and grassland which is heavily dressed with slurry, or which is mainly grazed, is unlikely to suffer from S deficiency.

Cutting removes up to 4 kg of S per tonne of dry matter, and this is most effectively replaced by periodic applications of 20–40 kg of S in the form of gypsum. More soluble forms of S can be used, but care is then needed to avoid too rapid uptake of S, which can affect the copper metabolism of the stock that are fed, particularly sheep. Thus while S-containing fertilisers are now readily available, their use should be based on the need for S, rather than as a routine application.

A number of other mineral elements, including copper, cobalt and zinc, are also needed in trace amounts for vigorous herbage growth, and for the feeding of healthy and productive animals. The supply and uptake of trace elements depends on many factors, including soil type (in particular pH), plant species, season and fertiliser usage. Most soils in the UK provide enough trace elements for optimum herbage growth, with deficiencies most likely on leached soils with low organic matter content; rather more commonly the herbage grown may not contain enough trace elements for the needs of the animals being fed. Diagnosis of deficiency is based either on visual symptoms (both plant and animal) or on plant analysis, because soil analysis is confounded by too many interfering factors to give reliable results. Overall, trace element levels can be increased by heavy applications of animal manures, and specific deficiencies can be made good by use of appropriate fertilisers, as with the application of cobalt sulphate or copper sulphate when herbage analysis shows it to be deficient in either cobalt or copper. However, manipulation of the trace element content of herbage is very imprecise (and excess application, in particular of copper, can result in toxicity to both plants and animals). Thus most cases of animal deficiencies are probably best dealt with by direct feeding of the appropriate mineral supplement.

Diseases of forage species

Permanent grassland generally contains many different plant species, which have developed a useful degree of resistance to a range of diseases. In contrast temporary grassland, often sown to a single variety or to a simple mixture of species, offers more opportunity for disease to develop; reductions in yield of 15 per cent due to ryegrass mosaic virus and of 35 per cent from crown rust disease have been recorded on badly infected sown swards. Fungal attack can also reduce the soluble carbohydrate content of cut forage, so making it less suitable for silage.

Diseases of grassland are most prevalent in the south and west of the country, though ryegrass mosaic virus is more generally distributed. In general, losses are lower than those noted above, but if disease does pose a problem it is advisable to sow more resistant varieties, to harvest earlier and more frequently – and to apply more N fertiliser. Under severe conditions, for example on grass to be cut for silage, a single fungicide spray is permitted.

Crops for hay

It is useful, then, to consider which, among the many factors that determine forage production, are the most important in relation to haymaking. Even with facilities available for barn-drying (p. 120), weather conditions in the UK are seldom ideal for making hay before the end of May, and probably later in the north. By this time the digestibility of most first-growth forage will have fallen below 65 D-value; because cutting for field-made hay is likely to be delayed until June, even with the most efficient field techniques the D-value of the hay that is made is unlikely to be above 60 (see Table 2.2). Thus there is advantage in making hay from fields sown to later-maturing varieties, or from fields which have been grazed until early May so as to remove many of the potential seedheads, both of which can provide forage of higher D-value for haymaking in early June. While the yield of hay from a sward that has been grazed will be lower than from an ungrazed sward this is perhaps not unwelcome, because there will be a better chance of making good hay from the lighter crop. Much hay is also made from longer-term leys and from permanent pasture, both of which, as has been noted, are likely to be lower yielding because they are generally less heavily fertilised than short-term leys.

In practice most of the forage that is harvested before midsummer is now conserved as silage, and most of the hay that is made is from lighter crops, cut later in the summer, which are the most likely to give a quality product with low losses. However, well-made hay is still recognised as a convenient and valuable feed, particularly for sheep and for young stock. This, coupled with the problems of safe disposal of silage effluent now being faced on many farms, and the more general availability of big-bale machines which can greatly speed up the hay harvest (p. 114), could lead to some increase in the amount of hay that is made. And, as is noted at the end of this chapter, hay is likely to be the preferred method of management on many species-rich old pastures.

Crops for silage

A much higher proportion of silage, in comparison with hay, is made from sown grassland well fertilised with N. The main species sown are perennial and Italian ryegrass which, as well as being of high D-value, also have higher contents of the water-soluble carbohydrates (WSC) needed for effective silage fermentation. Even the ryegrasses, though, if they are cut very early or very late in the season, or are grown with a high level of N input, may contain less than the target figure of 3 per cent WSC in the fresh forage (Table 4.6). Most other grass species, including cocksfoot

Table 4.6 Effect of level of application of nitrogen fertiliser on the water-soluble carbohydrate content of grass (per cent WSC in fresh grass)

Type of grass	*Low N (50 kg N/ha)*	*High N (100 kg N/ha)*
Italian ryegrass (RvP)	5.2	3.8
Early perennial ryegrass (S.24)	3.2	2.0
Late perennial ryegrass (S.23)	2.5	1.5

(Data: ADAS)

and timothy, and all the legume species, contain lower levels of WSC. Thus recent progress in breeding grass varieties with a higher WSC content could be important. Such varieties are likely to have a lower content of crude protein, which would make them more suitable for ensilage; they could also be higher yielding, because they synthesise more organic matter per unit of applied N fertiliser.

Ways of extending the period in spring during which forage can be harvested at the required level of D-value have already been noted. For, despite the very high rates of field-harvesting now possible, and the increasing employment by livestock farmers of specialist contractors using very high output equipment, there may still be advantage in producing a succession of crops of the same D-value over a period of two to three weeks, reducing the peak load on men and machines and making the overall operation less weather-dependent.

In the past one disadvantage of silage-making has been that it has not been a convenient or efficient way of conserving the relatively small quantities of grass, surplus to grazing needs, that become available at intervals during most grazing seasons. However, big-bale silage (p. 150) now makes it possible for quite small amounts of forage to be conserved quickly and efficiently; this is particularly useful later in the season when conditions become less suitable for haymaking, which has in the past been the main way of conserving such forage.

On many farms part of the area cut for silage is completely separated from the grazing area, either because the fields are too far from the farm buildings to allow easy access for grazing, or because they are planted to a crop specially grown for cutting. There is increasing interest in the latter option, and the following sections describe developments in two of the main crops now being grown for this purpose, forage maize and wholecrop cereals.

Forage maize

Maize was one of the crops brought back to Europe by Christopher Columbus, and was grown widely for grain in southern Europe. By the 1800s it had spread north, though now mainly harvested as a wholecrop for feeding to cattle, and by the late 1860s maize was being grown as a silage crop in France and Germany. However, the most rapid progress in ensilage was in North America, and by 1890, in Wisconsin alone, more than 2,500 silos had been built for the storage of maize silage. Experience with the crop showed the importance for efficient silage-making of a high dry-matter content in the crop, and by 1923 breeders in the USA had raised the dry matter in forage maize harvested for silage from less than 20 to more than 30 per cent.

Progress in Europe was much slower. The average dry-matter content of the wholecrop maize harvested in trials carried out at Wye College in 1901 was only 13 per cent; more than a third of the crop was lost in the ensiling process and the feeding value of the maize silage produced was estimated to be less than that of mangolds. E. J. Russell doubted if maize silage could compete with root crops, although he did recognise that the positive experience in the USA was probably due to the higher DM content in the crops grown there than in the UK.

It was not until the 1960s that hybrid maize varieties became available in Europe, mainly from extensive breeding programmes in France. These were earlier maturing (and so of higher dry-matter content) and higher yielding than previous varieties, and their use led to a rapid increase in the area of forage maize grown in mainland Europe, from less than 0.5 million hectares in 1965 to nearly 4 million hectares in 1994. Then, in the 1970s, and largely as a result of the enthusiasm of members of the Maize Development Association, trials with these new varieties were carried out in England, both at the NIAB and on commercial farms. However, the crop was still grown on only a limited scale, because the big year-to-year variations in both yield and dry-matter content that were found under practical conditions seemed to make it unsuitable for general use.

Following the introduction of milk quotas in 1984 many farmers sought to cut their feed costs by reducing the amounts of concentrates used and feeding more home-grown forages, in particular silages with a high energy value and intake potential. Despite the progress that had been made, this was often difficult to achieve on a regular basis with grass silage. Coupled with the increasing production costs and environmental concerns associated with high N usage on intensive grassland, this led to a renewed interest in maize silage – just at the time that a range of new, earlier-maturing varieties of forage maize began to be marketed, with dry-matter contents in the 30–40 per cent range, compared with the 20–25 per cent of the previous varieties. As a result there was a rapid increase in the area planted to forage maize in the UK. The crop was further encouraged by being included within the Arable Area Payment Scheme – although unfortunately the level of support was severely reduced in 1995 because the 85,000 hectares of forage maize grown in 1994 exceeded the area limit that had been set for the crop.

Areas suitable for forage maize

The rates of growth and development of maize depend critically on soil temperature during the early stages of growth, and later on ambient temperatures. Germination occurs when the soil temperature reaches 8–10°C; this usually happens between 25 April and 7 May, and determines the optimum planting date. During the growing season of May to October the maximum and minimum temperatures are used to calculate Ontario Heat Units (OHU), with 2,300 and 2,500 OHU needed for crops to reach 25 and 30 per cent dry-matter contents respectively. On this basis the Meteorological Office has produced maps showing the areas in the UK best suited to maize growing, and the expected crop maturity.

These maps, which provide a general description of areas suitable for growing maize, have been refined by Grainseed and the Coop de Pau to show which of the early-, medium- and late-maturing hybrids now available are best adapted to different areas of the country. In order to increase the chance of getting optimal crop maturity (in terms of both yield and dry-matter content) the earlier-maturing hybrids are recommended for the cooler northern regions, and later-maturing hybrids (which require less OHU to reach a given dry-matter content) for more favoured climatic areas of southern England.

In addition to choice of variety, site selection within a given locality is most important in relation to its effect on crop maturity and yield. Studies at Wye College have shown that high ground, exposed fields, north-facing chalk outcrops and cold heavy wet clays all delay crop maturity and reduce yield; crops on sheltered south-facing fields, which receive more OHU during the growing season, give the highest yields. A number of techniques for improving the local micro-climate are also being tested, including covering the drilled maize seeds with plastic film, and sowing the seed on the south side of ridges formed from east to west across the field. Equally as important is the production of a fine firm seedbed in which moisture has been retained. Soil compaction must also be avoided, particularly from equipment used to spread slurry in winter, because compaction can reduce rooting depth, with serious effects on yield (Table 4.7).

The mineral fertiliser requirements of forage maize, though they will depend on previous soil status and cropping, are in

Table 4.7 Effect of soil compaction on rooting depth and dry-matter yield of forage maize

Soil compaction	*Rooting depth (cm)*	*Relative dry-matter yield*
Low	130	100
Medium	80	94
High	35	73

(Data: Schroder, Netherlands)

general much lower than for grassland. Fertiliser is usually broadcast and worked into the soil at planting, with little response to more than 120:60:60 kg/hectare of N, P and K. Maize is very responsive to organic manures and, being a late-sown crop, many farmers are able to apply large quantities of slurry in the spring to fields that are to be planted to maize, so reducing the need for mineral fertiliser. However, although this may be environmentally preferable to winter application, there is little doubt that future legislation will more strictly control the timing, quantity and method of slurry application. The emphasis will then shift to efficient utilisation rather than disposal of slurry on fields to be planted to maize, using a regime such as that developed at NIRD in the late 1970s (Figure 4.6). This aims to minimise fertiliser use on the maize crop without applying excessive amounts of organic manures which may pollute water courses.

Because the main growth of maize does not start until mid-June, by which time some of the initial fertiliser may have been leached from the soil, there is a response to a later top-dressing with N. Conventional application of granules can cause scorching of the crop, and trials supported by the (now) Maize Growers Association have shown useful increases in yield from N fertiliser injected into the soil between the rows. This is quickly available to the plant roots at a time of rapid crop development, resulting in efficient nutrient utilisation without scorching, and with little risk of pollution (Table 4.8).

Weeds in forage maize have effectively been controlled by triazine herbicides, such as Atrazine, applied pre-emergence. However, herbicide resistance in weeds such as fat hen and black nightshade has begun to appear on fields which have grown maize consecutively for at least four years and which have had heavy dressings of organic manure. Under these circumstances a post-emergence combination of Lentagran as a contact herbicide to remove early-

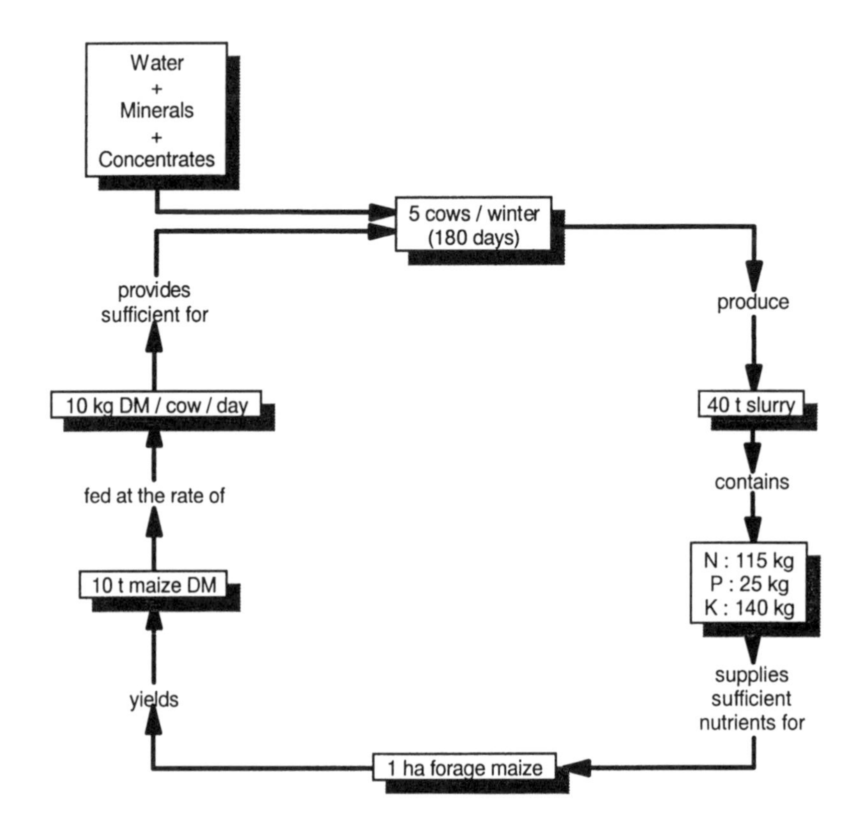

Figure 4.6 Recycling of mineral nutrients in the growing of forage maize
(Data: NIRD, Shinfield)

Table 4.8 Effect of injection of N fertiliser between the rows in late June on the yield and composition of forage maize

Treatment	*Dry-matter yield (tonnes/ha)*	*% dry-matter content of crop*	*% ears in crop*	*% CP in crop dry matter*
Control	16.8	32.3	67	7.2
60 kg N/ha	18.4	31.8	67	7.8
100 kg N/ha	18.0	32.3	67	7.8

(Data: CEDAR)

germinating weeds and Atrazine to provide longer-term cover is advised. Another contact herbicide, Bromoxynil, which has been used for many years in mainland Europe, has recently been approved for use on maize crops in the UK.

However, there is increasing environmental concern over the long-term use of residual herbicides (and not only with maize) and research is in hand on mechanical methods of weed control, and on genetic engineering to develop varieties of maize (and of other crops) which are resistant to the main contact herbicides.

Regular assessment of forage maize varieties, carried out by the NIAB with the active support of the Maize Growers Association and the Maize Agents Association, shows a mean yield from current varieties of about 14 tonnes of dry matter per hectare, with D-values ranging from 70 to 75, and metabolisable energy contents from 10.8 to 11.5 MJ of ME per kg of dry matter. The huge current maize-breeding programmes in several EU countries, in particular in France, are certain to provide still higher-yielding varieties, as well as the earlier varieties that will allow further northward cultivation of this crop in the UK.

Wholecrop cereals

Despite the considerable interest in the use of forage maize for silage production, at the end of the 1970s maize was still a rather unreliable forage crop in many parts of the UK, because sufficiently early-maturing varieties were not then available. In 1978 the Silsoe Working Party therefore recommended that research should be carried out on the potential use for silage production of cereal crops which would be better adapted to UK conditions, in particular wheat, barley and oats. This was not a new concept; for many centuries the feeding of wholecrops of cereals as forage had been practised on a small scale in regions of low summer rainfall, where grassland had not provided a reliable source of fodder. Interest had been renewed with the introduction of ensilage at the end of the last century, and crops of cereals, and of oats and vetches, were grown and harvested for silage.

These crops were generally cut at a fairly immature, high-moisture stage, at which they contained enough soluble carbohydrate to allow them to be ensiled in conventional bunkers and clamps (p. 143). Losses, however, particularly in silage effluent, had in practice been very high, and the first priority in the new study of the ensiling of wholecrop silages was to apply the principles, now in general use for grass silage, of field-wilting, short-chopping, rapid filling and sealing of the silo, and of good management to prevent wastage during feeding out. Initial studies showed considerably reduced losses when wholecrop cereals were ensiled in this way.

However, this work also confirmed the results of Danish research, which had shown that silage made from immature cereal crops seldom gave as good animal performance as grass silage, and then only with low-producing stock. Furthermore, the yield of wholecrop cereals cut at an immature stage was much lower than the yield from the same crop allowed to grow to a more mature stage. The Working Party thus advised that research should be carried out on methods of conserving wholecrop cereals harvested at a relatively mature stage.

It was quickly found that mature crops were likely to suffer high losses if they were ensiled by conventional methods, and this led to the development of the methods of preservation by alkali (high pH), described on p. 33. The success of these new methods, and in particular the high efficiency of preservation when urea was applied at 4 per cent of the dry weight of the wholecrop, stimulated new research on the yield, moisture content and digestibility of a range of different cereal crops, harvested over a wide range of maturities. This research showed the very high yield potential of current cereal varieties, and confirmed the considerable yield penalty suffered when a cereal crop was harvested at an immature stage, appropriate for storage in a clamp or bunker silo, as against the more mature stage that was now practical with the addition of urea (Figure 4.7). It was also found that, unlike the digestibility of

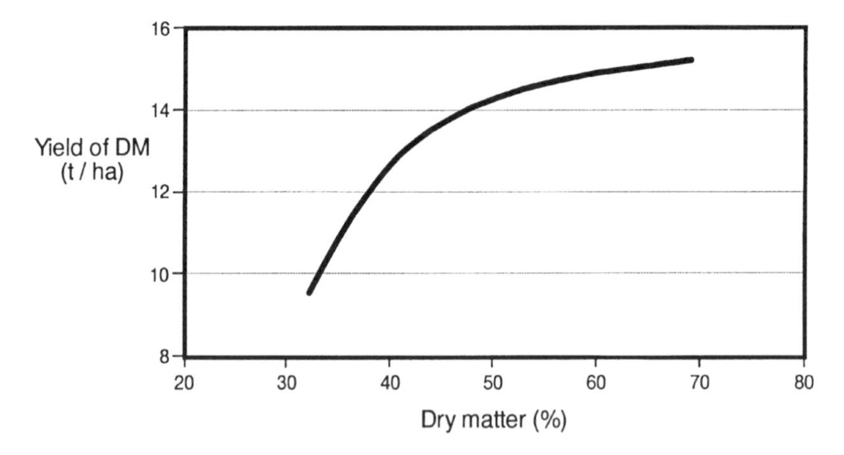

Figure 4.7 The yield of dry matter of wholecrop wheat (variety Fortress), and changes in its dry-matter content, as it becomes more mature (Data: Wye College)

grass crops (see Figure 3.2), there is only a small decrease in the digestibility of wholecrop cereal during a period of up to 40 days after first heading, because the increasing proportion of (highly-digestible) grain in the crop largely compensates for the decrease in digestibility of the straw as it becomes more lignified. Furthermore, the research also showed that preservation with both urea and caustic soda slightly increased the digestibility of the silage that was produced (Figure 4.8).

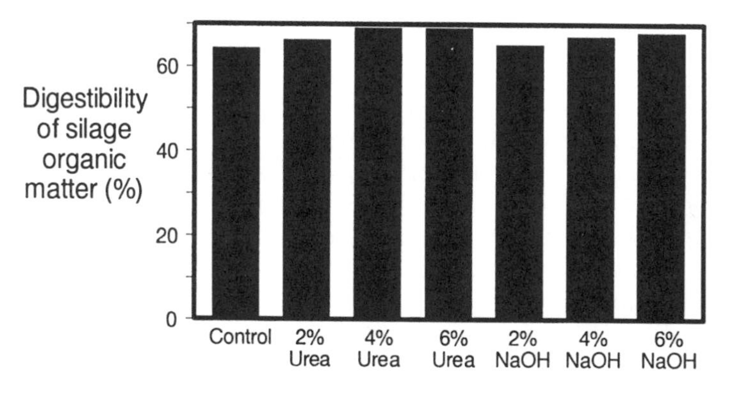

Figure 4.8 The digestibility of silage made from wholecrop wheat, either alone (control), or with additions of urea or sodium hydroxide (Data: IGER, Hurley)

These results stimulated considerable farm interest in wholecrop cereal silages, an interest further encouraged by wholecrop being included under the crops eligible for support under the Arable Area Payment Scheme introduced in 1993 (effectively, a subsidy of £20 per tonne of dry matter stored as wholecrop silage). Thus although, as described in the previous section, the introduction of the new early-maturing varieties of maize during the 1980s led to a big increase in the cultivation of forage maize, production of wholecrop silage also increased – to the extent that the Maize Growers Association set up a Wholecrop Cereals Group to monitor and support developments with this crop. It had perhaps been expected that wholecrop would be adopted mainly in the north of the country, where conditions are less suitable for the cultivation of forage maize; however, a survey made by the group in 1990 on 53 farms on which wholecrop silage had been made found that most of the farms were in the southern half of England (with nine of them in Devon, suggesting an enthusiast in that county!). Most of the farmers reported that wholecrop silage was cheaper to produce than grass silage (even before receipt of area support for the wholecrop), and experience with feeding was generally encouraging (p. 199). However, most of the farmers also concluded that more research was needed on crop production and harvesting techniques, on improving the reliability of the preservation process and on optimising feeding regimes.

In practice the agronomy of wholecrop cereal is very similar to that of the same crop grown for grain, in terms of sowing date, seed rate and fertiliser inputs, and the relative yields of dry matter from different species and varieties of cereals harvested as wholecrop are similar to the corresponding grain yields. In particular wheat and oats tend to outyield barley (Table 4.9), and yields from winter-sown crops are higher than from spring-sown crops. Dry-matter yields ranging from 7 to 12 tonnes per hectare were recorded in the farm survey noted above. However, a number of operators now leave a 15 cm straw stubble at harvest, having concluded that the loss in yield is more than compensated for by a higher D-value, because the harvested crop contains less straw; a similar benefit may result from the application of a straw-shortener to the growing crop.

Table 4.9 Field-scale yields of wholecrop cereals harvested at 50–60 per cent dry-matter content

Cereal type	*Yield range, tonnes dry matter/ha*
Winter wheat	9–15
Winter barley	7–12
Winter oats	7–15
Triticale	8–18

(Data: ADAS)

Certainly the nitrogen input to wholecrop, at 120–150 kg N per hectare, is much lower than the 400 kg N per hectare needed to produce a similar dry-matter yield from grass; more of the N input to the cereal can be supplied by animal slurry; and grass also has to be harvested several times in the season, as against the single harvest for the cereal crop. Wholecrop, though, is not necessarily as 'environmentally friendly' and 'energy efficient' as these figures might imply, because the urea used in the ensilage of the cereal wholecrop is equivalent to a further input of up to 250 kg N per hectare; much of this N will be excreted, mainly in urine, but also as gaseous ammonia, when the cereal silage is fed. Thus every effort must be made to reduce the amount of urea required, by ensuring that it is applied uniformly to the cut crop, by rapid sealing of the silo to prevent loss of ammonia, and by the use of urease to increase the activity of the urea that is added (p. 34).

There is also interest in the possibility of reducing the input of fertiliser N by growing the cereal in combination with a legume (shades of the oats and vetches of earlier days!). In what has been termed 'bi-cropping', the cereal seed is slot-seeded in the autumn into an established white clover/grass sward. The clover provides much of the N requirement of the growing cereal crop until it is harvested as wholecrop in the following July; the sward is then grazed hard from mid-August to mid-October to stimulate further growth of clover before it is again slot-seeded with cereal seed. Work at the IGER has shown that, although the yield of wholecrop grown in this way is lower than from conventionally grown wholecrop, this may be compensated for by the considerably lower input costs (Table 4.10). The silage made also has a higher nutritive value than straight wholecrop because it contains a proportion of high D-value clover.

Table 4.10 Comparison of yields of wholecrop cereals from conventional and bi-cropping treatments

	Conventional	*Bi-cropping*	*Difference*
Silage yield (tonnes dry matter per hectare)	16.9	11.2	5.7
Cost (£/ha) of fertiliser and other chemicals	227	54	173

(Data: IGER, 1994)

As has been noted, a particular attraction of wholecrop cereals is their suitability for areas where forage maize cannot be grown, for reasons either of soil type or of climate. On a mainly grass farm an area sown to spring cereal can be used as a way of utilising large amounts of slurry. The crop will also be harvested earlier than if it had been grown for grain, so allowing more time for preparation for the next cereal crop, or for direct reseeding with a grass/clover ley. Wholecrop also offers flexibility of use, because the decision whether to cut a given cereal crop for silage, or to harvest it later for grain, need not be taken until after most of the 'grass' silage has been made, and the fodder situation for the following winter is known. By early July, in a dry summer, it may also be apparent that the farm faces a shortage of grazing in the immediate future; the decision can then be taken, at fairly short notice, to cut and store a quantity of wholecrop cereal silage, probably in the form of big bales, to be used as a 'buffer' feed over the following weeks (p. 223). Alkaline wholecrop silage also provides a useful neutralising effect in the rumen when it is fed in combination with wet, low pH grass silage.

Kale and crop by-products for silage

Kale, sugar-beet tops and vegetable wastes can also be ensiled, with potentially the largest by-product being the tops and crowns from the 200,000 hectares of sugar-beet that are grown in the UK. With modern harvesting equipment these can be collected relatively free from soil contamination, but effluent loss when tops are ensiled is high because of their generally low dry-matter content, in the 14 to 20 per cent range. However, most beet-growing farms have little requirement for conserved forage because they keep few ruminant livestock, while the task of ensiling sugar-beet tops can make a considerable demand on labour at a time when harvesting the roots is the number one priority. Thus, without considerable changes in farming systems, and in the economics of livestock feeding, there is unlikely to be any significant utilisation of beet tops.

In the immediate post-war years there was much interest in the growing of forage brassica crops, and in particular in the ensiling of kale for winter feeding. However kale, like beet-tops, produces large amounts of effluent when it is stored in a conventional bunker silo, and the ensiling of kale largely ceased in the 1960s. As a result very few brassica crops are now grown as animal feed, with the area in England and Wales having declined from 189,000 hectares in 1960 to about 12,000 hectares today – and most of that grazed rather than cut.

Recent work at IGER, stimulated by Devon farmer Ray Patey, has re-examined the potential role of kale silage, but ensiled now in plastic-sealed bales, either bagged or wrapped (p. 152), because this allows much better control of both fermentation and effluent loss than when kale is ensiled in a bunker silo. Maincrop kale, planted directly after a grass sward has been harvested for silage in late May, and well fertilised with farmyard manure and slurry, has given autumn yields in excess of 9 tonnes of dry matter per hectare; alternatively, a 12-week catch-crop of kale has yielded 5–6 tonnes of dry matter, but of higher digestibility than the maincrop. These yields are lower than those for forage maize, but kale has a higher crude protein content than maize, and is also less sensitive to altitude and soil type.

Unlike the earlier direct-cutting, the kale in the work at IGER has been cut with a mower-conditioner, leaving a well-set-up windrow on a 74 mm stubble. After wilting for 24–48 hours the crop was baled with a variable-chamber baler to give a dense centre to the bales, and the bales wrapped within 12 hours. With favourable weather conditions the dry-matter content of harvested kale can reach 20 per cent, but in these preliminary experiments, and despite the wilting, it was closer to 15 per cent; the kale ensiled well, however, with a pH of 4.2. The resulting bales weighed close to 800 kg and in order to minimise effluent loss – measured at 46 litres per bale – they were stored in a single layer. The silage has been readily eaten by stock; though animal production data are not yet available, baled kale silage, fed out with an automatic bale shredder (p. 174), may offer a potentially useful new conservation crop.

CROPS FOR HIGH-TEMPERATURE DRYING

In the immediate post-war period small grass-drying units were installed on many livestock farms, and larger drying units were set up by the (then) Milk Marketing Board to dry green crops for farmers on a cooperative basis. However, the first priority on most of these farms was to grow enough grass for grazing; as a result crops for drying were often available for less than 80 days during the season, which proved quite uneconomic. On the relatively few farms which have continued to operate high-temperature drying (p. 16) the key to economic success has been the provision of a steady flow of high-quality green forage, so as to keep the expensive harvesting and drying equipment operating over as long a season as possible – often on more than 200 days during the year.

Programmes to secure a succession of crops for drying have been greatly helped by information on crop yields and digestibility, such as that described earlier. The quality objectives of high protein content and high D-value cannot always be found in every crop that is harvested, but each crop that is dried should either be of high digestibility, such as ryegrass in mid-May (though still ensuring that it contains more than the 15 per cent of crude protein needed to attract the EU subsidy!) or of high protein content, such as lucerne and red clover (which are also grown for their soil fertility building properties).

Clearly, to secure such a steady flow of forage crops for drying requires a high level of field management. The expertise needed is that of the arable farmer, with advantage to large-scale operation because much the same level of skill is required to plan and manage the cropping on 100 hectares of land as on 1,000 hectares. Thus while successful grass-drying does continue on some mainly grassland farms, most of the production is from operations in which crops for drying are grown as alternative cash crops in what is effectively an arable farming enterprise.

FORAGE CONSERVATION/PLANT AND ANIMAL CONSERVATION AND DIVERSITY

Forage conservation during the 1970s and early 1980s was almost solely concerned with growing, harvesting and storing as high yields as possible of high feeding-value crops. Permanent pasture continued to be ploughed up and replaced by leys, often sown to only a single species; increasing amounts of fertilisers, particularly N, were applied to grassland; and with the shift from hay to silage more fields were cut earlier, and more frequently, than in the past. This resulted in a huge reduction in the diversity of both plant and animal species within the grassland areas of the UK. This was perhaps accepted as long as 'food output' was the main priority; but as the risk of food shortage has been replaced by apparent food surplus, there has been increasing public concern about the possible longer-term environmental consequences of such intensive farming systems.

The response was slow (thus grants for removal of hedges were still being paid in the early 1980s), but a range of schemes has now been introduced that seek to address the conflict between maximising production and maintaining species diversity in the countryside – generally by compensating the farmer for at least part of the income lost if land is farmed less intensively so as to increase species diversity and protect landscapes. Thus payments for the protection of Sites of Special Scientific Interest, introduced with the 1981 Wildlife and Countryside Act, were initially extended in the Environmentally Sensitive Areas programme and, more recently, in the Countryside Stewardship Scheme, which aims to re-create some of the diversity that has been lost, for example on chalk downlands and old meadows.

Many fields still contain viable seeds of many different 'weed' species, and these can be reinforced by seeding with wild flower seed mixtures which are now marketed for a variety of soil types. But the key to the establishment of a botanically diverse sward is management – in particular in reducing the amounts of fertiliser applied in order to prevent grass dominance and to encourage 'low-fertility' species; delaying cutting the sward until the plants have set seed; and conserving the cut crop as hay rather than as silage, because haymaking encourages greater seed dispersal. Late cutting is also recognised as essential to the survival of ground-nesting birds and small mammals, whose numbers have been greatly reduced through intensive grassland management.

All this is a far cry from the high-output forage production systems described earlier in this chapter – and leads on to the growing debate as to how far society should, or will be prepared to compensate farmers for the lower output of environmentally sensitive farming, and for the costs of re-creating cherished landscapes – and, most critically, whether existing environmental schemes, which to date cover only a small fraction of total farmland, and attract only a tiny part of total agricultural support funding, should be extended more generally to the whole of the countryside of the United Kingdom.

Undoubtedly the choice, and the management, of crops for forage conservation will vary greatly. The farmer feeding store cattle on permanent grassland, most of whose hay will come from surplus grass not needed for grazing in June and July, will have quite different priorities to the operator of a high-temperature dryer aiming for crops to be cut in sequence every day from April to November. On most farms, however, cutting and grazing are managed as a single overall operation, aimed at providing both quantity and quality in both grazed and cut forage. Most of this forage is still likely to come from swards of perennial grasses and legumes, and most of the forage that is cut will continue to be conserved as silage.

But the problems posed by the underlying pattern of growth of perennial forages, shown in Figure 4.1, remain. It is for this reason that we have given what may be considered undue importance in this chapter to the potential of annual forage crops, such as maize and wholecrop cereals, with the aim of making the provision of winter feed for farm livestock more secure.

CHAPTER 5

MOWING AND SWATH TREATMENT

To make hay or wilted silage water must be removed from the cut crop in the field as rapidly as possible, and with minimum loss of dry matter. To achieve this the mowing, conditioning and other operations that are applied must be suited to the forage species and yield of the standing crop, the stage of growth at which it is cut and the expected weather conditions. Heavy crops make drying more difficult, and losses during the drying process are likely to be greater the drier the crop has to be before it is fit to be removed from the field. In particular, as drying proceeds the leaves in the cut crop become brittle and tend to break off whenever the crop is moved. These small fragments are difficult to recover so that, even when the most efficient haymaking techniques are used under fine weather conditions, up to 10 per cent of the crop dry matter is likely to be lost.

Dry-matter losses are also greater under wet conditions because the crop may have to be moved several times before it is dry enough to lift; rain falling on the cut crop also leaches out soluble (digestible) nutrients, with the result that the digestibility of the remaining crop falls (see, for example, Table 2.2). This happens each time the crop is rewetted, with a modest rewetting of a partly dry crop often causing more damage than when heavier rain falls on a crop that has just been cut. Much more moisture has to be removed from crops to be conserved as hay than for silage; as a result, crops for hay are much more likely to be affected by rain or heavy dew and field losses are generally much higher with haymaking than with silage.

To achieve the overall aim of getting a rapid rate of moisture loss with minimal dry-matter loss, the principles of crop drying described in Chapter 2 must be fully exploited; in particular some physical treatment should be applied to the crop soon after it is cut in order to get the fastest possible drying rate under the prevailing weather conditions. Such treatment is most important when a conventional cutter-bar mower is being used, because the swath formed behind this type of mower, if left intact, loses moisture only very slowly. (To avoid any risk of confusion in terminology, and although not all authorities would agree, in the following discussion we have used 'swath' to refer to the uniform, fairly flat layer of grass left behind a cutter-bar and some disc or drum mowers, while 'windrow' describes the grass from one or more such swaths gathered into a row across the field. Alternatively the cut crop from either a swath or a windrow can be 'spread' randomly across the whole area of the field that has been cut.)

The basic requirements of mowing and conditioning equipment

Mowing is a critical operation. The mower must cut the standing crop cleanly to the required stubble height, and then 'pass it on' to the next stage of field handling. Depending on the weather conditions this second stage generally aims either to set the crop up into windrows, evenly distributed across the field, and with a regular cross-section and open structure to allow moisture-removing air to pass through the crop, or to spread it loosely over the surface area of the field, so as to get the maximum rate of drying by the sun, but possibly more at risk to rainfall. For both hay and silage the crop is then finally set up in windrows of uniform density and of a width that can be cleanly picked up by a baler or forage harvester; balers require a windrow of particularly uniform shape and density across their width.

To achieve a high output the mower must be capable of operating continuously at a high forward speed, and without blockage, in a wide range of crops, including laid crops with a heavy and dense bottom growth. For reasonably efficient operation less than a quarter of the operating time should be taken up by turning and transit between fields or by delay for adjustments. The cutting mechanism should also be designed to prevent stones and other objects being thrown upwards – though it is not always possible to avoid the ejection of missiles.

The mower must also be able to cut the required area of crop at the optimum stage of maturity (digestibility); 1.5 to 2.0 ha per hour is a reasonable target on many farms, but larger operations, in particular those harvested by contractors, achieve much higher rates. In the past this target was not always achieved, because many of the mowers used had a cutting width of less than 1.8 m, which would have required a forward speed of 15 km per hour; however, most current mowers have a cutting width above 2 m, which can achieve the required work-rate at lower and more acceptable speeds. The cut crop left by these wider machines must later be gathered into windrows so as to pass between the tractor wheels, and be of a width to match the subsequent handling and harvesting equipment.

The power requirement of the mower must be well within the working capacity of the tractor. This may seem self-evident – but it is essential not to underestimate the power needed at full forward speed in a heavy crop, and the standard of mowing and recovery of crop are frequently disappointing when an underpowered tractor is used, particularly on sloping ground.

Types of mowing equipment

Some requirements of the 'ideal' machine have already been noted. The aim here is thus to examine how the equipment now available matches up to the demands of different conservation systems.

Reciprocating finger-bar mowers were once the main method of cutting, but they are now seldom used because of the considerable time needed for knife-sharpening and maintenance, in particular on stony ground. They can give very effective cutting in well-standing crops with little bottom growth, leaving a clean and even stubble and little or no fragmentation of the crop, so that dry-matter loss attributable to mowing is very low. In contrast, even well-maintained mowers can give problems in wet and heavy laid crops, with 'bunching' along the swath following a succession of blockages, and considerable loss from uncut stubble.

Finger-bar mowers are simple in operation, lightweight and cheap to produce, and have a low power requirement of about 2 hp per metre width of cut. Forward speed is generally within the range of 3–8 km per hour; thus with a 1.5 m cutting width overall output can range from 0.6 ha per hour in a clean standing crop to as little as 0.2 ha per hour in a badly laid crop – both, it should be noted, considerably lower than the target 1.5 to 2.0 ha per hour indicated above. Much higher rates of work were possible with the *fingerless double-knife mower*, which had a multiple scissors-like action that gave very clean cutting. However, these machines were more expensive, required frequent knife-sharpening and are now seldom used.

In the early days of silage-making most grass was cut with a finger-bar mower and brought to the silo by buckrake (the Paterson system), but in the 1950s the *flail harvester*, originally developed in New Zealand, was introduced. This machine cuts the crop by the shearing action of flails rotating at up to 1,000 rpm on a horizontal rotor arm. The powerful airflow set up by the flails is also used to deliver the cut crop, via a chute, to a transport trailer. However, the flail harvester had the disadvantage that crops were direct cut, so that there was no wilting in the field; the powerful suction action of the flails, combined with some scalping action, also tended to contaminate the silage with soil.

With the recognition of the importance of wilting green crops before they are ensiled, the operations of cutting and lifting began to be separated, and flail harvesters were modified to allow a two-stage process, with the flail mechanism operated first as a *flail mower*, returning the cut crop in a windrow onto the ground and then, after a period of wilting, being used to lift the wilted crop for transport to the silo. Flail mowers were also used for cutting crops for hay. This proved to be a rather inefficient method of mowing, however, and most forage crops are now cut with either a *rotary disc mower* (Plate 5.1) or a *rotary drum mower* (Plate 5.2), using the horizontal slicing action of freely swinging knives, fitted to the periphery of the disc or drum, and travelling with a linear speed of about 80 m per second. In their simplest form both types of machine leave the cut crop in an open swath, but most are now fitted with deflector plates which set the cut crop up in windrows (Plate 5.3). Many mowers are also directly coupled to conditioning units (see p. 98).

Disc and drum mowers have a much greater capacity than cutter-bar mowers, but they do require between four and eight times as much power – typically 15 hp per metre cut at the pto, and double this with a mower that has badly worn or blunt blades, or with the cutting angle or height badly adjusted. Knife maintenance with both disc and drum mowers is a simple operation, and up to 40 ha of crop can be cut under favourable conditions before a blade change is necessary – though more frequent changes may be needed on flinty land or with incorrect blade setting. Given adequate power these mowers work efficiently and without blockage even in heavy, laid crops. However, particular attention needs to be given to the adjustment of the height of cut and to the fore and aft setting of the cutter unit so as to avoid 'scalping' the sward. This can cause excessive blade wear, and produces an uneven stubble and a 'mane' of crop left between the cutting discs, which can significantly reduce the subsequent rate of regrowth of the crop. Properly operated, both types of mower can give a very clean cut with little fragmentation of the herbage 'stubble' remaining on the field.

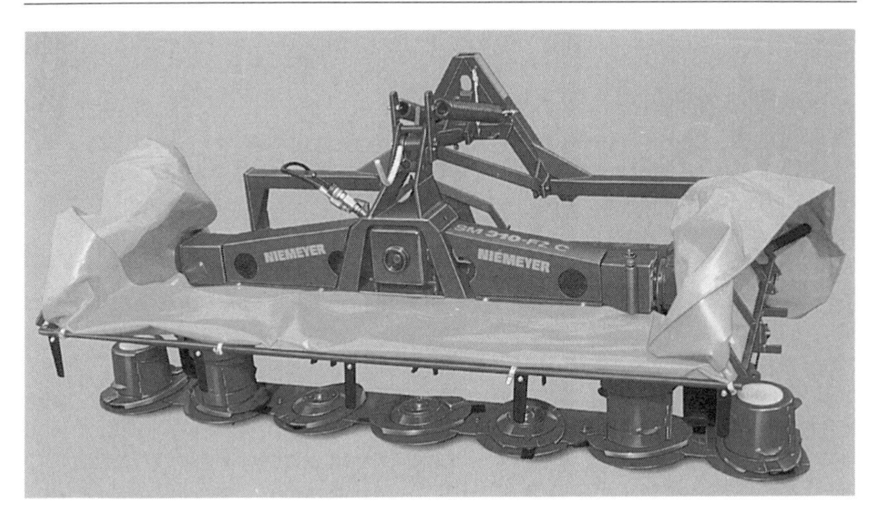

Plate 5.1 Multi-disc mower (Niemeyer, UK)

Plate 5.2 Multi-drum mower (Niemeyer, UK)

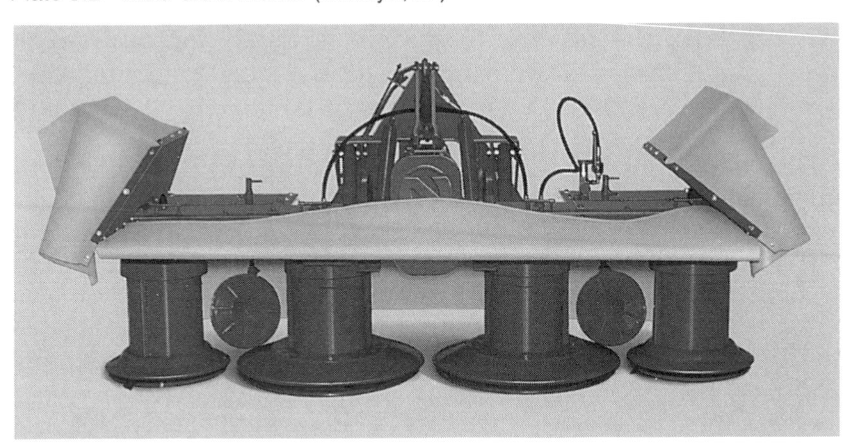

Care must be taken to avoid double-cutting, which can occur if the mower passes over already cut crop, and to prevent the tractor running over already cut crop, by adjusting the deflector plates fitted at the rear of the mower so as to form a windrow within the width of the tractor wheels. The width of cut of these mowers has steadily increased, however, and combinations of front- and rear-mounted mowers can give overall mowing widths above 7 m, and cutting rates up to 10 ha an hour (Plate 5.4); special mowing and field handling routines are then needed to exploit these machines fully.

Plate 5.3 Windrows of grass cut by eight-disc rotary mower with rear deflector plates (Westmac Ltd)

In general, rotary mowers can be operated at a forward speed of up to 16 km per hour, with a continuous operating rate of 10–13 km per hour – a cutting rate of between 1.0 and 1.5 ha per hour per metre width of cut, depending on crop and ground conditions, and the skill of the operator. They give little physical conditioning, though, and unless the cut crop is opened up by adjustment of the deflector plates it may be left with the thickest and wettest parts of the crop on the soil at the bottom of the windrow and the more easily dried parts at the top.

Treatment after mowing

Mowing the crop removes the growing plant from its source of water, and the crop then begins to lose moisture through evaporation. However, the amounts of water that have to be lost are

Plate 5.4 Coupled front and rear mowers (Kverneland Kidd Ltd)

considerable – in the case of haymaking as much as four times the final weight of hay – and with a finger-bar mower the rate of loss from the cut swath may be very slow unless it is quickly opened up to allow drying from the action of wind and sun. This is done using a range of machines, generally termed tedders or rakes, and whether or not some primary conditioning is combined with the cutting operation (p. 98), tedding or raking is generally applied so as to tease open, mix and turn the windrows as drying proceeds. When good drying conditions prevail the most rapid drying is achieved when the cut crop is more widely spread over the field. This is done using a multi-purpose tedder or rake, which can open up the crop from swaths and windrows and spread it thinly over the whole ground area (Plate 5.5) to give the maximum rate of drying by the sun; collect the spread crop back into windrows when there is a serious risk from rain and then open it up again when drying conditions improve; and finally collect the spread crop into windrows ready for harvesting (Plate 5.6).

A wide range of machines is now available, and selection must take account of the different tasks the machine will be required to carry out in the particular hay or silage system being used. Correct setting and operation are critical, and an important facility is the ability to adjust the tines to suit the particular crop and stage of

Plate 5.5 Tedder spreading crop from windrows (Niemeyer, UK)

Plate 5.6 Rotary rake gathering spread crop into a large windrow (Barclays Bank plc/CLAAS (UK))

drying. Except for the largest machines tedders are generally mounted on the tractor three-point linkage; single and double rotor machines work at rates of 1.2 to 2.8 ha per hour, with four-rotor machines covering up to 4 ha per hour.

Crop conditioning

Tedding greatly increases the rate of moisture loss from the cut crop and, when correctly applied, gives uniform drying which reduces the risk of damp patches when hay is being made. Tedding by itself does little to speed up the rate of moisture loss from the thicker, stemmy, parts of the crop, however. The most effective way of doing this is to 'condition' the crop by applying a physical treatment as soon as possible after it is cut, while it is still turgid. The power requirement is increased considerably if the operation of mowing is combined with one of the conditioning operations, described below, when between 20 and 30 hp is required per metre working width on the mower, if output and performance are not to suffer. The conditioning mechanism is also likely to be more susceptible to damage from stones and metal trash than the mower alone, and design features aimed at reducing this risk, coupled with ease of maintenance and repair, deserve special attention.

Conditioning equipment

Conditioning the cut crop operates in two ways, first by breaking up the waxy layer on the plant surfaces which retains moisture within the living plant, and second by splitting open the plant stems so as to produce a bigger surface area for evaporation. Earlier conditioning machines operated with two main mechanisms: *crushing*, in which the cut crop passes under pressure between plain or fluted rollers, which flatten but do not shorten the stems, and have little effect on the leaves; and *crimping*, which uses corrugated rollers to crush the stems and bend them at 50–100 mm intervals, as well as causing some bruising of the leaves.

Crushers and crimpers are also used to 'set up' the crop as soon as possible after it is cut so as to produce a loose windrow, through which air can pass to give rapid drying. These machines are, however, sensitive to damage by stones, which can 'burr' the crimper bars, and this can result in excessive treatment of the crop. They were developed in the USA, mainly for use in crops of lucerne and red clover, and have proved less effective for conditioning the heavy grass crops more typical of the forages harvested in the UK. Thus other conditioning systems have been developed for grass crops.

Most of the crops ensiled during the 1950s and early 1960s were cut by a flail harvester and loaded directly into a trailer for transport to the silo – in the process often producing the wet, badly fermented silage that made the system so unpopular. Thus the *flail mower* was developed by replacing the vertical delivery chute on the flail harvester with baffle plates which directed the cut crop back as a windrow on top of the cut stubble (p. 93). These cut crops dried much more rapidly than if a cutter-bar mower had been used, because the flail cutting mechanism, operating against the baffle plates, caused a considerable amount of splitting and shredding of the crop which greatly speeded up the rate of moisture loss.

Modifying the type of flail and the rotor speed allowed some control over the degree of laceration and fragmentation of the cut crop; thus double-edged swan-neck flails, operating against a shear-bar at a rotor speed of 1,200 rpm, produced the short chop required for silage, while heavy-duty flails operating at 800 rpm gave some laceration, but without breaking the crop into smaller fragments, and were therefore more suited to haymaking. Crops cut by flail mower gave much more rapid drying than similar crops cut with a reciprocating blade mower.

Flail mowers worked well in laid crops, particularly if they were operated with the 'lay' of the crop and, provided enough tractor power was available, they did not suffer from blocking and bunching. Output varied from 0.4 to 1.2 ha per hour for a 1.5 m cut, and up to 1.5 ha per hour with the wider 1.8 m cut. However, they needed considerable power, and under most conditions at least 45 hp was needed to operate at a forward speed of 8 km per hour. Flail mowing is also a fairly robust treatment, which tends to smear the plant sap over the surfaces of the cut crop. This makes the plant fragments sticky, producing a rather compact mass from which soluble material can be lost even with light rain. Flail mowing also produces a proportion of very small plant fragments that are easily lost during wilting.

Thus although flail mowers were widely used during the early stages of the shift from hay- to silage-making that began in the early 1970s (see Figure 1.2), they were then steadily replaced by disc and drum mowers, which give more efficient cutting with lower power requirement. However, these mowers give less conditioning action than flail mowers, and a considerable research programme, directed by the late Gordon Shepperson, was set up at NIAE, Silsoe, to develop specialised conditioning equipment. This established a novel principle of crop conditioning, in which the cut crop, either picked up from the field or fed directly from the cutting mechanism on the mower, enters a rapidly rotating rotor(s) fitted with spokes or brushes which throw the crop at high speed against a closely fitting hood. The slip between the crop and these elements abrades the waxy cuticle on the surface of the crop, while the spokes also penetrate and treat the whole mass of the crop, so giving a very uniform treatment. Spokes comprising V-shaped steel elements and plastic elements were tested, while in a later design nylon brush units were fitted to two intermeshing rotors (Plate 5.7), with the resilient but slightly flexible plastic effectively breaking the protective waxy layer on the plant surfaces.

These principles have been incorporated in a range of commercial conditioning machines. Some of these have plastic elements, others use either swinging or fixed metal spikes, while others use metal rollers intermeshing with plastic brushes. With all these machines the dimensions and speed of the rotor and the location

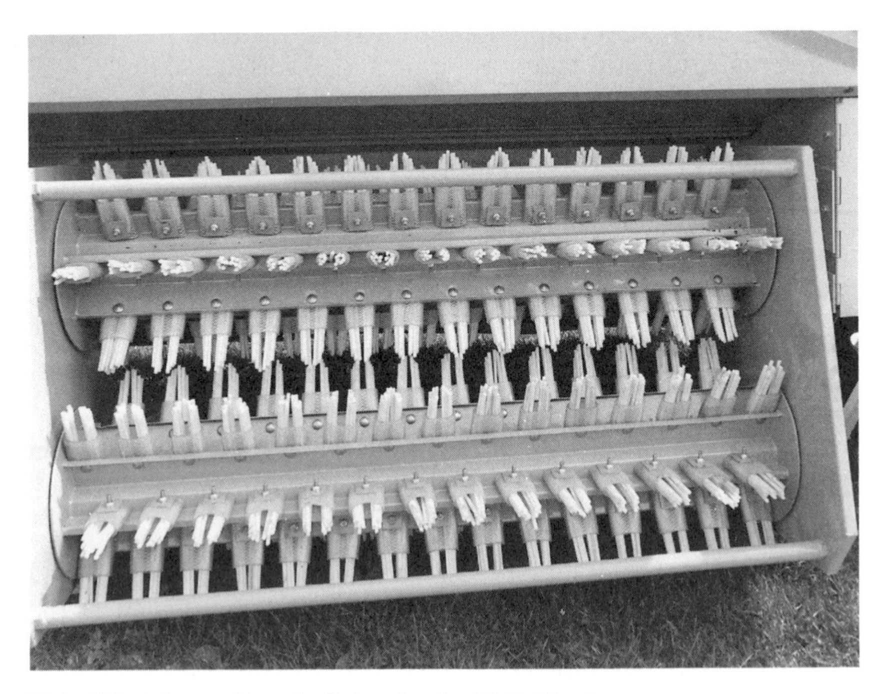

Plate 5.7 Intermeshing plastic brush rotor (NIAE, Silsoe)

and shape of the spikes or brushes are designed to achieve a high conditioning effect without breaking the crop into small pieces, while ensuring that the crop disengages from the rotor without wrapping or blocking. Furthermore, by adjusting the deflector plates at the rear of the unit the conditioned crop can either be set up in a regularly shaped windrow, ready to be picked up by a forage harvester after a period of wilting, or spread more thinly to get maximum drying by the sun. Properly operated, conditioners can reduce the time needed to dry a mown crop to the 25 per cent dry-matter content needed for many silage systems, to less than two-thirds of that with conventional tedding equipment, and can also greatly speed up the drying of crops for hay.

Combined mowing and conditioning equipment

Where rapid wilting is required with good labour economy there is much advantage in combining the processes of cutting and conditioning the crop into a single machine. Conditioning can then be carried out as an integral part of the cutting action (as in the operation of flail mowers, already described).

The first such equipment tested in the UK was the combination of finger-bar mower and crimper/crusher, already noted, that had been successfully used in the USA. However, this equipment did not work well with heavy grass crops, and all the mower-conditioners now used in the UK are based on drum or disc mowers, from which the cut crop is fed directly into the conditioning unit (as in Figure 5.1). Commercial mower-conditioners all employ a high-speed horizontal rotor(s), fitted with spikes, tines or brushes or ribs, which takes the cut crop directly from the mower and physically opens up the plant structure without producing small plant fragments. The conditioned crop is then returned to the field, either as an open windrow (Plate 5.8) or spread more thinly, by adjusting the rear deflector plates.

Operating mowing and conditioning equipment

The ideal mower-conditioner should have a work rate at least equal to that of the mower operating alone, and with no greater risk of blockage or breakdown; the addition of a conditioning rotor, however, can double the power requirement compared with cutting alone, and adequate power, of the order of 35 hp per metre of cutting width, is essential. Thus one of the largest mower-conditioners (Plate 5.9), with a cutting width of 9 m, has a power

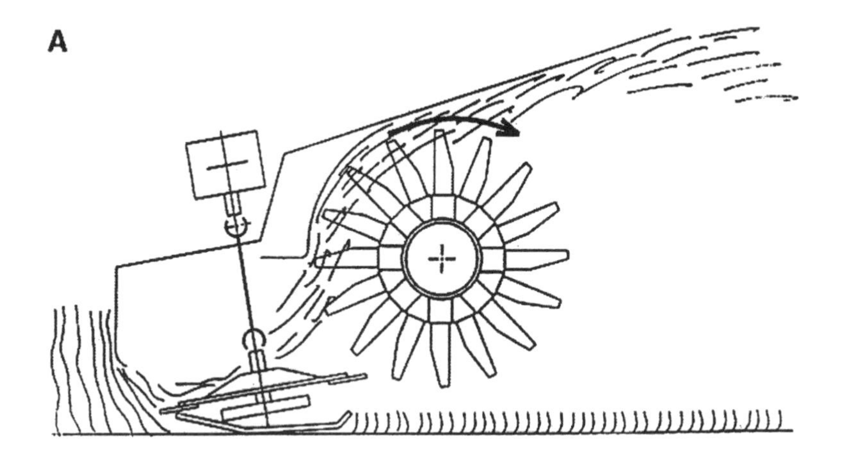

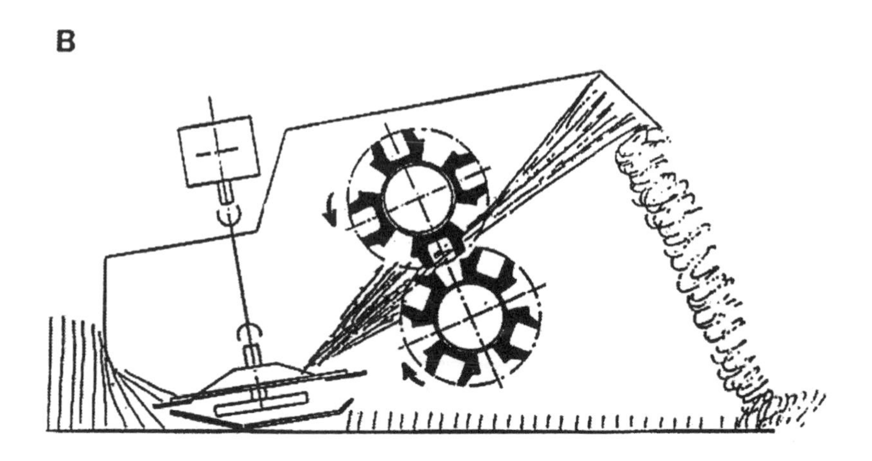

Figure 5.1 Design of mower-conditioner units
A. Single rotor with flexible polyethylene or nylon fingers
B. Intermeshing ribbed rubber rollers (Westmac UK)

requirement of about 200 hp and can cut and condition up to 7 ha of crop per hour.

It is important that the structural strength of the crop should be retained so that the windrow does not collapse when the treated crop is returned to the ground; the stems should also be more severely treated than the leaves, which naturally dry more rapidly; and damage to the plant tissues should not be so severe as to make them susceptible to leaching by rain, or to loss of small

Plate 5.8 Windrows set up by mower-conditioner (Agridry Ltd)

Plate 5.9 Nine-metre wide self-propelled mower-conditioner (CLAAS (UK))

fragments. It should also be possible to adjust the severity of conditioning to the type and yield of crop. The windrow produced should retain an even and open structure, if possible with the slower-drying stems exposed on the surface; and if no subsequent treatment of the crop is to be carried out before it is lifted, for example with medium-wilted silage, the windrow formed behind the machine should be of a size and shape that can be picked up directly by the forage harvester without an intervening operation.

These features may now seem obvious but, as we have noted, they only became clear following the detailed studies at Silsoe, and there is still much to learn about the optimum design and mode of operation of this type of equipment.

A number of general points apply to the operation of all types of equipment. Thus heavy rolling early in the season of all the fields that are to be cut is essential, to prevent breakages and hold-ups by stones and other objects; this is particularly important when flail and drum/disc mowers are to be used. Attention to preventing the formation of lumps of grass in the mown swath will pay dividends throughout the whole operation, particularly in haymaking, because lumps dry more slowly and this either delays final baling, or produces damp patches within the bales which later mould. Even with silage-making a 'lumpy' crop can seriously slow down the harvesting operation.

Experimental evidence shows that the total annual yield from a sward is generally higher the closer to the ground the crop is cut. It is difficult to put this into practice on a field scale because of the difficulty of cutting close to the ground without 'scalping' parts of the sward, which can severely reduce the rate of regrowth and also risks contaminating silage with soil. Under most practical conditions a cutting height of about 40 mm above soil level is advised – although there are experienced operators who are firmly convinced that the cutting height should be no lower than 100 mm!

The importance has already been noted of applying the first tedding or conditioning treatment to the crop either at the time of cutting or as soon as possible after it has been mown – indeed, delay not only reduces the effectiveness of conditioning equipment but may also make it difficult to operate. The windrows produced by an effective conditioner have an open structure which aids drying and, unless the crop is to be made into hay or high dry-matter silage, they may not require further treatment before they are lifted. However, as already noted, many crops will require secondary treatment (tedding), and this must be carefully graded according to the type of crop and the stage to which it has dried.

Haymaking

The area of crop that is cut for hay should be matched to the available mowing, tedding (conditioning) and baling capacity, so that the crop on each field or part of a field can be baled as soon as it reaches the required degree of dryness (see Table 2.1). To do this it may be necessary to ted the crop at intervals during the day, and it is generally advisable to collect the partly dried hay into compact rows overnight so as to minimise take-up of moisture from dew. These windrows are then moved sideways or, under good weather conditions, spread over the whole land area as soon as the ground between the rows has dried out the following morning. Large dense windrows should be avoided while the crop is still green and moist, as these dry very slowly.

With haymaking, once the crop in the windrows has been dried down to the required moisture content it should be baled as quickly as possible. However, if rain does fall before the crop is baled, the windrows should be quickly opened up into the original rows as soon as rain ceases, otherwise subsequent redrying will be very slow. The type of windrow presented to the pick-up reel on the baler is also very important. The maximum rate of baling is achieved when the baler is operating at a low forward speed with a heavy and dense windrow, producing bales with as few as six 'wads' per bale. However, final drying out in field stacks of this type of bale, particularly if weather conditions are poor, can be very slow, and better-quality bales containing a larger number of 'wads', and which dry out and store better, may be made by operating at a higher forward speed with a lighter windrow.

Crop loss and drying rate

Many factors affect the rate at which the cut crop dries in the field, and the extent of dry-matter loss during the process. In general, the lower the moisture content that is required in the crop before it is removed from the field (see Table 2.1), the higher will be the risk of losses. The previous sections have indicated how machines may be selected and operated so as to maximise drying rate while minimising crop loss. Few formal comparisons have been made between the range of machines now available, but a number of general observations can be made.

Losses in the form of uncut stubble are of most importance when the field is only being used for conservation, for if the regrowth is to be grazed (see Figure 4.2) much of the crop that is not cut will be eaten later by the grazing animals. Rotary mowers operated at the correct speed can cut as cleanly as reciprocating cutter-bar mowers, but stubble losses may be high if the knives are blunt or if the mower is operated at too high a forward speed in relation to the tractor power. However, rotary mowers tend to give lower losses than finger-bar mowers in laid crops or crops with a heavy bottom growth.

Losses from fragmentation differ widely between machines, but in general are higher when flail and rotary mowers are used than with a cutter-bar. Against this must be set the timeliness of the operation, with the newer high-performance mower-conditioners reducing field exposure time by at least two days, compared with traditional mowing and tedding; in turn, the latter combination can give a further two-day advantage compared with leaving the mown crop untouched in the swath.

Increasing the number of secondary treatments can speed up the drying rate under favourable conditions, but can also make the crop more vulnerable to bad weather. If the prevailing weather conditions indicate that several field treatments may be needed, therefore, it is advisable to apply the most severe treatment first, for this can often reduce the need for later treatments by at least one pass through the crop. Such possible savings in field treatment should always be considered when evaluating the potential advantages of alternative mowing and conditioning equipment.

These considerations are most significant in the case of haymaking, which requires the crop moisture content in the field to be reduced to between 15 and 25 per cent (except in the rare cases where storage drying is available). Silage, for which crops can be harvested at a much higher moisture content, generally in the 65–75 per cent range (25–35 per cent dry-matter content), poses less problem, and field losses are in general much lower, because the leaf fraction does not become brittle until the crop moisture content has fallen below 60 per cent.

The crop has been mown, conditioned and tedded. The following chapters describe the further operations needed to store the crop as hay or silage.

CHAPTER 6
HAYMAKING

The previous chapter examined systems of mowing and conditioning grass and forage crops so as to encourage rapid and uniform rates of drying in the field. As Table 2.1 showed, most methods of haymaking, other than barn-drying with heated air, require the moisture content of the cut crop to be reduced below 30 per cent before the hay can be removed from the field. Once the required moisture content has been reached haymaking then becomes largely a materials-handling problem, with each subsequent movement of the hay adding to the cost without improving the feeding value of the product.

BALES AND BALERS

Most hay is still stored in standard bales, 360 × 450 × 900 mm in size, holding between 14 and 23 kg of hay, sufficient to provide the daily hay requirement of four to ten dairy cows. Most hay is also still fed by hand, and this size of bale is convenient to manhandle and feed, especially when small amounts of hay have to be fed at some distance from the barn. Baled hay needs only half the storage space of loose hay, and the density and shape of bales greatly increase the capacity of both transport vehicles and stores. The main drawback of the standard bale is that it is too small to be fully suited to individual mechanical handling, yet too large to be moved by conveyors in the same way as grain. As a result the movement of bales from field to store and from store to livestock nearly always requires hand work, and no method of handling yet developed gives a free-flowing system comparable with the mechanised systems now available for silage.

Manual handling of standard bales can be minimised by moving them in unit loads by a tractor foreloader from groups of bales formed in the field directly behind the baler. Hand work is also reduced if hay is stored in large round or rectangular bales (p. 114).

Standard balers

Most balers are of slicing-ram design (Plate 6.1), in which the reciprocating ram, which presses the hay into the bale chamber, carries a knife which cuts through the flow of crop at the end of each stroke, producing compact bales, in which the 'slices' are cleanly separated. Bale density depends on the type and moisture content of the crop and can also be altered over a wide range, from 80 to 220 kg/m^3, by adjusting the spring-loaded pressure plates. On most balers the length of the bales can be altered over a range from one and a half to twice the bale width, using a simple star wheel and tripping arm operated by the flow of crop through the chamber.

There are minor variations between different makes of baler, but the optimum requirements are common to all. Thus the pick-up should be wide enough to deal with a heavy crop set up in a field windrow at least 1.5 m wide. A combination of narrowly spaced pick-up tines, a freely floating pick-up head and a spring-loaded crop guide appears to be the most effective. The crop must then pass in a clean unobstructed flow from the pick-up to the feeder chamber and thence to the bale chamber. Here efficient packers which can be easily adjusted are essential if well-shaped and rugged bales are to be formed.

Plate 6.1 Pick-up of slicing-ram baler (John Deere, Ltd)

There has been a progressive increase in the speed of operation of the baler ram, to over 90 strokes per minute at standard pto speed, and this has led to smoother running, especially when the ram is mounted on sealed ball-bearing rollers. Uninterrupted operation is also assisted by the heavy flywheels that are now fitted, which smooth out surges caused by variations in windrow size and density, and so reduce the peak power requirement. Modern baler knotters can operate with a range of twines, though minor adjustments to twine tension may be needed.

Bale-grouping devices are often towed directly behind the baler, and it is essential that these do not impose an uneven load on the bale chamber. A rapid and easy method of converting the baler from the 'transport' to the 'working' position, without the tractor driver having to leave his seat, is also a great advantage. Comprehensive out-of-season maintenance is essential, as well as daily servicing, including adjustment to the knotter and needle assemblies.

The practical operation of a baler to produce regular well-shaped bales of even weight can only be learned in the field, but a number of general points are worth noting. Thus the baler may need to be adjusted at intervals during the day to cope with variations in both the type and the moisture content of the crop; otherwise bales may vary – from very heavy in a crop of moist, leafy hay to very light in a stemmy open crop of very dry hay – with bursting of strings at one extreme and collapse of bales at the other. Adjustment will almost always be needed at sunset, whether or not dew has started to fall, and also when hay is being baled on field headlands near hedges or woods.

Hay is generally baled when it still contains between 20 and 35 per cent of moisture. At the most common moisture content of 25–30 per cent, bale density will be about 160 kg/m^3, giving a standard bale weight of about 23 kg, but this can increase to over 210 kg/m^3 and a bale weight of 32 kg at higher moisture contents. Heavier bales may be produced if an extended delivery chute, connected to the rear of the bale chamber, increases the back-pressure.

If hay is to be barn-dried it will probably be baled at a moisture content above 30 per cent, and the tension in the baler must then be reduced to avoid producing bales of too high density: a good guide is that it should be possible to thrust half a hand into the side of each bale after it has been released from the chamber. At higher moisture contents it may also be advisable to reduce the bale length to 700–800 mm in order to keep the bale weight below about 32 kg. This shortening has the further advantage that the bales that are formed are then more likely to remain intact within the strings when they are handled after drying.

Another factor in bale stability is the number of wads (strokes) per bale, which depends on the rate at which the crop is being picked up and the ram speed. This is seldom critical with heavy crops, but bales made from a fairly light crop at high forward speed and with less than about 20 wads can become rather loose and require careful handling from the field. Bales that are to be barn-dried should contain fewer wads, with the ideal bales, containing only 8–10 wads, being produced when a large windrow of crop is being picked up at a high forward speed.

Baler output

Baler output can vary greatly depending on the crop and weather conditions. However, general guidance may be useful in deciding on the area of each crop that should be cut at any one time, and on the capacity of the ancillary equipment that will be needed to move the baled hay from field to store. Under practical conditions the output of most standard balers ranges from 3.5 up to 10 tonnes per hour, with a common on-farm rate of 6–7 tonnes (250–300 bales) per hour, equivalent to the hay from one hectare of crop. While it may be possible to bale at a faster rate, this rate is already greater than most bale-handling systems can deal with – a sizeable team of men, handling say 45 bales per man-hour, would be needed to clear bales being produced at this rate.

Handling standard bales

The mechanisation of bale handling is done in five stages: grouping, loading on to transport, transport, unloading and loading into store. In practice there is advantage in forming the bales into small stacks for a holding period, either in the field or under temporary cover, before they are moved into store. This allows evaporation of residual moisture without the hay heating up, while giving useful weather protection. By introducing this intermediate holding stage between baling and final storage, handling is done in two distinct stages, which can be of advantage when only a small team of workers is available.

The shape and size of the stacks of bales formed in the field must be planned as an integral part of the overall handling system, yet without restricting the throughput of the baler. The dimensions of the handling equipment must also be directly related to the field size and topography, and to the storage building size and layout. Thus poor accessibility at the buildings may limit the choice of equipment, while storage capacity may be inefficiently utilised if the bales are brought in in unit loads which do not match the dimensions of the buildings, leaving space too small for further unit loads.

Handling single bales

It is, of course, not necessary to group bales before they are collected, and single bales can be loaded onto trailers from the rows on the field where they were dropped from the baler. However, hand-loading is now seldom done and a tractor-mounted swinging arm or an extended chute and thrower attached to the rear of the baler can be used to load the bales directly into a towed trailer – although hand-stacking on the trailer does slow down as the day progresses. The alternative, of tumble-loading the bales into the trailer, is sometimes used, but this reduces trailer capacity, and the overall loading rate can then fall below 200 bales per man hour.

Handling bales in groups

Single bales as left by the baler are easily damaged by rainfall, and they also interfere with the regrowth of the crop if they remain on the field for more than a few days. For these reasons, and because, as noted above, stacks of bales can continue to lose moisture in the field, bales are now often dropped in groups across the field, using equipment operating directly behind the baler, and are then built into stacks, generally by hand. These stacks, which provide the first stage in the overall handling system, can then be left to allow some further drying out in the field. If there is risk of rain they can be partly covered by small plastic sheets, with the sheets being removed at intervals to prevent condensation. This practice is more common in the generally more difficult climate of the north of England and Scotland.

The bales must first be collected into groups without interfering with the output of the baler. Various forms of collector, towed behind the baler, have been developed for this purpose, many of them by farmer inventors. Manned sledges are seldom used, because the work-rate falls as the operator tires, and unmanned random collectors are more commonly employed, the simplest type consisting of a metal cradle, towed behind the baler, which collects batches of up to 20 bales which are then released in loose heaps in rows across the field (Plate 6.2).

However, the bales left by random collectors have to be stacked by hand, and automatic bale accumulators (Plate 6.3), towed behind the baler, have much advantage. The most common design arranges the bales into a single flat layer of eight bales, which is then released onto the ground. Several of these groups can then be formed by a tractor foreloader into a stack of 40–100 bales, which is then left to lose further moisture until it is ready to move into store. Bales from these field stacks can be loaded manually on to a trailer at over 300 bales per man hour, but are more efficiently handled as complete units from field to store using a tractor foreloader. Where the initial grouping is of a single layer of bales these can be impaled from above for lifting on to a trailer (Plate 6.4), while stacks of more than one layer are handled by a side-gripping loader. The rate at which bales can be loaded in this way depends very much on trailer size and on the layout of the field stacks. However, with good organisation an impaler loader, operated by one man, can load up to 500 bales per hour, or more if the trailer can take four units of eight bales in each layer.

Special-purpose carriers, which collect and transport pre-formed stacks of bales directly from the field and so eliminate the need for loading on to a trailer, are also used, including combinations of front- and rear-mounted buckrakes; capacity, and safety in transit, can be increased by fitting the carrier with hydraulically squeezed sides. Larger loads can be carried by a rough-terrain loader, but

Plate 6.2 Random bale collector. Sledges of this type collect 15 to 20 bales, which are then dropped in rows across the field

Plate 6.3 Automatic bale sledge. Models are available which form bales into flat layers of eight or ten which are then automatically discharged in a compact form ready to be picked up by an impaler loader

Plate 6.4 An impaler loader can handle bales in either flat-eight or flat-ten formation, preferably loading directly onto a large flat-bed trailer

transit distance can quickly limit the output of tractor-mounted systems, which falls off rapidly when loads of less than 50 bales have to be carried more than a kilometre.

All these systems work best when the stacks of bales formed in the field have vertical and straight sides, so that they fit neatly within the extended frames of the carrier. Stacks that have been poorly constructed, or which have distorted by settling, pose serious handling problems – most often found with small stacks of bales that have been handled several times. To prevent this, equipment has been developed which bands each block of bales together immediately after it has been formed behind the baler and before it is released on to the field. Using this equipment one man can collect and cart up to 3,000 bales per day, a rate comparing well with any alternative unit-handling system. However, this system has not been widely adopted, at least in part because of its high cost and complexity.

Loading into store

Most attention has been given to clearing hay from the field as rapidly as possible so as to minimise the risk of weather damage, and mechanisation at the store is still often limited to the use of an elevator to move the bales from trailer level into the store. This involves double manual handling, however, and labour use at the store can then be as high as in the combined field and transport operation.

This has emphasised the advantages of handling hay in unit loads directly from the field into the store, in particular with unit dimensions that match the bay-size in the barn, and using a loader able to stack bales at least fourteen layers high.

Large bales

While most of the hay that is made is still stored in standard bales, an increasing proportion is now handled in much larger round or rectangular bales, weighing between 200 and 500 kg. The first effective large baler was designed, and built, in 1967 by two Gloucestershire farmers, David Craig and the late Pat Murray, and was in commercial production for a number of years. This produced rectangular bales, 1.5 × 1.5 × 2.4 m, each made up of a series of small bundles of hay, arranged in parallel across the bale and held together by three heavy-duty polypropylene ties. This 'bundle' formation produced a rather low bale density, of

80–100 kg/m³, which made the bales open to penetration by rain, so that it was advisable to move them under cover as quickly as possible. However, the low density did make them very suitable for both natural and forced-air ventilation, and they were used in many barn hay-drying installations during the 1970s. Correct swath preparation was important in order to produce regularly shaped bales which would stand being handled. Handling was best done with a tractor foreloader fitted with a 'gripper' which closed hydraulically on the sides of the bales.

Baling rates of 6–9 tonnes per hour were typical, with transport by trailer and loading into store at 3 tonnes per man hour, and the bales formed very stable stacks when they were stored with the flat sides against a retaining wall. Feeding out was best done with a tractor foreloader, the 'bundles' of hay within the bale making it easy to distribute to stock.

However, in the 1980s these rectangular bales for hay were largely superseded by large round bales, which are also now widely used for the storage of straw and silage (p. 150). A number of different round balers have been developed, but in general they operate on one of two main systems:

- Fixed chamber round balers (Plate 6.5) produce bales of a fixed diameter. Hay, picked up from the field windrow, is fed into the bale chamber and subjected to an increasing rolling action

Plate 6.5 Fixed chamber round baler. The fixed rollers forming the bale can be seen (Greenland UK Ltd)

as it begins to come into contact with the fixed bale-forming mechanism on the periphery of the chamber. This operates with rollers, chain and slats or belts in different models, and these produce a bale containing a spiral of compressed crop, with the outer layers more densely packed than the core. The finished bales are then secured either with a spiral of twine around the circumference, or more commonly with a layer of plastic net, which also gives some protection against rain if the bales are left in the field. Bale size is determined by the diameter of the chamber, and bale density can be controlled by the tension on the release mechanism.

- Variable chamber round balers (Plate 6.6) have a bale-forming mechanism which expands as the hay is fed into the chamber. The hay is thus subject to continuous pressure and as a result the density of the bales is greater and more uniform than with the fixed chamber machines; bale size can again be altered by operating the tying mechanism when the required diameter is reached.

Current machines of both types produce bales ranging in size from 0.9 m diameter × 1.2 m width up to 1.8 m diameter × 1.5 m width. Bale density with dry hay is about 150 kg/m^3, with different sized bales weighing between 0.3 and 0.5 tonnes. These machines are relatively simple to operate, but skill and experience, in particular in setting up the windrows of hay that are to be picked up in the field, are essential to ensure that uniform and stable bales are produced. Work-rates, which depend on both crop yield and bale size, range from 6–10 tonnes of hay per hour – with bale tying and ejection taking up to 40 per cent of the operating time with some machines.

However, despite these high work-rates large round balers are not ideal for haymaking, because the high density and large size of the bales restrict the escape of the heat and residual moisture from still-moist patches of hay within the bale, so that there is more risk of moulding than with standard bales. To minimise this risk hay that is to be round-baled should be field-dried to a lower moisture content than is normal with standard bales, and the bales should always be allowed to dry out in the field until their moisture content is below 20 per cent. In theory the bales with a lower-density core produced by fixed chamber machines should be less at risk, but these tend to pack down in store, again reducing overall air movement. As a result all types of round hay bale are less suited to artificial ventilation than rectangular bales. It has also proved difficult to distribute a preservative chemical (p. 14) uniformly in

Plate 6.6 Variable chamber round baler. The forming belt can be seen
(New Holland UK Ltd)

round bales, and work at NIAE, Silsoe, has shown that two to three times more ammonium propionate may be needed to preserve hay at 25 per cent moisture content in round bales than is needed for hay at 30 per cent moisture content in standard bales.

Round bales are generally handled by tractor foreloader attachments and transported either by trailer or bale-carrier, using systems mainly developed for handling straw. They can be stacked by tractor foreloader (Plate 6.7), but their shape limits the efficient use of storage space, and stacks more than a few bales high are not very stable.

For all these reasons large round bales have not been widely adopted for haymaking, though some farmers, particularly in the north, who use round balers for making silage, have found them well adapted to making hay from relatively light regrowth crops in mid-summer. The bales are then often left to dry by natural ventilation in the field, with the plastic wrapping of the curved surface of the bales, now possible with several types of baler, giving useful weather protection.

There is thus considerable interest in the potential for haymaking

Plate 6.7 Stacking large round bales (Merlo (UK))

of a new generation of balers, operating on the principle of a scaled-up standard baler, which produce large rectangular bales that are more stable and easier to handle than round bales (Plate 6.8). The largest such bales, of the order of 1.2 × 1.3 m in cross-section and 2.5 m in length, and holding over half a tonne of hay, allow very rapid transport of hay from field to store. However, as with round bales, it is essential that the moisture content of the hay is reduced below 20 per cent before it is stored, and this will often require a period of further drying out in the field after the hay has been baled, unless some system of artificial ventilation is available.

Balers of this capacity are large and expensive machines, and are more likely to be operated by a contractor than by the individual farmer. This has encouraged the development of machines which produce rectangular bales of intermediate size – of the

Plate 6.8 Large rectangular baler (New Holland UK Ltd)

Plate 6.9 Loading large rectangular bales onto a transport trailer (New Holland UK Ltd)

order of 0.5 × 0.8 × 2.0 m, and weighing up to 0.25 tonne, and well adapted to many larger farms.

Both large and intermediate sized rectangular bales provide rugged packages that can be loaded from field to trailer (Plate 6.9), from trailer to store, and from store to feeding area by both conventional and high-lift tractor foreloaders.

BARN HAY-DRYING

Previous editions of this book gave considerable attention to the potential role of barn-drying in reducing the weather dependence of haymaking – a technique in which one of the original authors, the late Gordon Shepperson, was the national expert. However, as we noted on p. 12, with the progressive replacement, since the mid-1970s, of hay by silage as the main method of forage conservation, interest in barn hay-drying has steadily diminished, and very few drying units continue to operate.

We consider it unlikely that this situation will change, in particular in light of the high labour demand of the system and the increasing pressure on labour costs, and we have thus decided not to include here the detailed earlier discussion of barn-drying. We would, however, make one caveat, that barn-drying might regain importance if new environmental priority is placed on the re-establishment of traditional species-rich meadows (p. 88). In that case late cutting of hay, which allows the different species, in particular non-grass species, to set and shed seed, might be more widely adopted. Small, specialised barn-drying units might again then provide a useful specialist method of haymaking.

CHAPTER 7

SILAGE-MAKING

As silage time approaches two main considerations come into focus – the desired stage of growth for cutting the crop and the likely weather conditions.

All the obvious things will (or should) have been put in hand: machinery checked and ready, or the contractor booked; buildings and silos prepared; plastic sheeting purchased; and the chosen silage additive to hand, either for planned use or as a precaution against bad weather. Ideally the time at which the crop reaches the optimum yield and D-value will coincide with good weather, so that it can be rapidly wilted to the level of dry-matter content needed to make good silage.

A notable development since the last edition of this book has been in the increasing number of livestock farmers who now employ contractors to make their silage. The use of contractors for silage-making brings with it both plus and minus factors. The main advantage is that the larger team and higher-output equipment used by the contractor will generally do the job much faster, and often more efficiently, than would be possible with the available farm labour and equipment; many farmers, as the time comes to replace their existing silage-making tackle, are likely to follow this route. On the downside, though, there may be pressure to get on when conditions are less than ideal. Thus, while good relations between contractor and customer will usually achieve a fair compromise even in catchy weather, not all silage will be made under optimum conditions.

However, whether farm or contractors' equipment is used, the farmer still needs a clear understanding of the key factors contributing to successful silage-making. There is now much more general agreement than perhaps in the past about these factors, and this has underpinned the continuing expansion in silage production. The factors can be divided into those concerned with the field work, and those that relate to storage.

In the field:

- Cut the crop at the optimum combination of yield and stage of growth (D-value), based on the considerations discussed in Chapter 3.
- Decide on a minimum dry-matter content of 25 per cent, to be aimed for except under an exceptional run of wet weather, and make frequent use of weather forecasts in deciding how much crop to cut each day. Ensure that the harvester is set to give the optimum chop-length for the dry-matter content of the crop (Table 7.1).
- Eliminate all chance of soil contamination at every stage of harvesting and loading into the silo.
- Apply an additive whenever necessary to ensure good preservation and to produce silage of high feeding value.

Table 7.1 Maximum chop-length for crops to be ensiled at different dry-matter contents

Crop dry-matter content (%)	*Maximum chop-length (mm)*
Under 20	200
20–25	130
25–30	80
Above 30	25

At the silo:

- Fill the silo by a method that mimimises air movement, heating and oxidation within the crop.
- Consolidate the crop during filling as necessary, cover as soon as filling stops each day, and completely seal the silo against the entry of air as soon as the filling of each section of the silo is completed.
- Cover the surface sealing sheet to hold it in contact with the silage below, and check at intervals throughout the storage period to ensure that the sheet remains undamaged and in close contact with the silage.
- Ensure that any effluent that does escape from the silo is effectively collected and disposed of safely.

Some operators may consider it possible to ignore one or two of these recommendations. But we believe that the approach should be the same as the principle of the three-legged stool – if any leg is missing or faulty the whole thing is likely to fail!

HARVESTING THE CROP

Most of the silage fed in the United Kingdom is still made from grasses and clovers, and these crops are considered first in the following sections. While many of the same principles also apply to the ensiling of forage maize and wholecrop cereals, of course, those crops do present their own particular problems and so are examined separately.

Stage of growth at cutting

Enough has already been said to underline the vital importance of this factor, and the results of research showing the characteristic D-value/yield relationships for the main forage varieties can be of great value in helping grassland farmers to decide when they should aim to cut each crop. Yet failure to cut forage crops at the right stage is still widespread, and the temptation to wait a few days for 'a little more bulk' can seriously reduce crop quality. It cannot be overstressed that the stage of growth of the crop at cutting will have more influence on the eventual feeding value of the product than any other factor under the farmer's control – hence the high priority we give to it.

Dry-matter content: the 'wilting' debate

Most forage crops contain less than 20 per cent of dry matter at the time they are cut, and much of the silage made in the 1950s and early 1960s also contained less than 20 per cent of dry matter because it was made from crops that had been cut and directly loaded by a flail forage harvester. Research then began to show the advantages of increasing the dry-matter content of forage crops before they are ensiled, in terms of the reduced weight of crop to be carted from the field, the improved fermentation in the silo, the higher feeding value of the silage produced and the reduced effluent loss from the silo.

This led increasingly to the separation of the operations of cutting and lifting, so that the crop can be wilted in the field before it is taken to the silo. At first only a limited amount of wilting was applied because, with the ensiling methods then used, crops that had been heavily wilted tended to overheat in the silo. However, by the early 1970s this problem had been largely overcome through the adoption of the sealed-silo techniques, described below, together with the use of silage additives, and as a result the average dry-matter content of silages made in the UK began to increase towards the 25 per cent level.

Then, by the late 1970s, and following reports of the success of the high dry-matter silage system being used in the Netherlands, some advisors began to recommend much more extensive wilting, to a crop dry-matter content in the 30–35 per cent range. It became evident, though, that this system might be more practicable in Holland, where most silos held less than 50 tonnes of silage and could be filled and sealed in a single day, than in the United Kingdom, where many farm silos, which held more than 500 tonnes of silage, were filled over a period of a week and so were more susceptible to the weather. Much of the grass to be cut for conservation is at the required D-value in the last two weeks of May and in early June, when the dry-matter content of the crop is still below 20 per cent, and when weather records make clear how difficult it is to guarantee rapid wilting. Thus to require a high degree of wilting seemed to make it less likely that each crop would be cut at the optimum D-value, particularly in the higher-rainfall grass-growing regions of the country, where two consecutive fine days are generally needed to increase crop dry matter to above 30 per cent.

Thus while some farms did adopt high dry-matter silage-making, the most widely accepted advice during the 1980s was that each crop should be conditioned as soon as possible after it was cut, and then wilted for up to 24 hours, with the aim of getting a minimum dry-matter content of 22 per cent, and an optimum dry-matter content of 25 per cent, before it was lifted and brought in to the silo.

However, the recent development of equipment which, under reasonable weather conditions, can greatly increase the speed and reliability of field wilting, has led to renewed interest in the ensiling of higher dry-matter grassland crops. In particular the latest types of mower-conditioners, described on p. 101, coupled with the concept of spreading the conditioned crop over the whole surface of the field, rather than leaving it in windrows, can give much faster rates of wilting. Experimental comparisons have shown crop dry matters in the 30–40 per cent range after 24 hours field-wilting, compared with 25 per cent with more conventional equipment.

When such heavily wilted crops are loaded into the silo, however, it is even more important to ensure rapid consolidation and exclusion of air, and this places particular importance on short-chopping before they are ensiled; backed up by a strict regime of daily covering of the silo, this can give much more effective control of heating and fermentation than was possible in the past.

Certainly the ensilage of high dry-matter forage crops is now a more practical proposition than was the case 20 years ago. Yet our own view remains, that to insist that every crop must be wilted to a dry-matter content above, say, 30 per cent before it is ensiled must reduce the chance of each crop being cut and stored at its optimum D-value. Equally, though, harvesting crops at below 20 per cent dry-matter content is likely both to reduce the chance of a good silage fermentation and to increase the volume of silage effluent produced. Thus we would continue to advocate what we believe remains the majority view, that there is great advantage in wilting the conditioned crop for up to 24 hours after it is cut, with the aim of raising the dry-matter content as close as possible to 25 per cent before it is brought in to the silo. If the weather has not allowed this degree of wilting, the crop should still be brought in after 24 hours, rather than leaving it in the field at risk of further loss and deterioration, while at the same time disrupting the overall harvesting operation. But in that case a silage additive (see Chapter 2) will almost certainly be required, and additional precautions must be taken to prevent effluent pollution. On the other hand, when conditions for field-wilting are ideal, the crop may well be above 25 per cent dry-matter content after 24 hours' wilting, and an additive may again be needed to control heating in the silo. However, this higher degree of wilting is not essential, either to ensure a good fermentation, or to exploit the higher nutritive value of wilted crops (p. 51), and if the rate of moisture loss in the field is very rapid it may be advisable to limit the degree of wilting by shortening the time the crop remains on the field.

Clamp or bunker silos, or sealed big bales, are the preferred methods of ensilage because they provide an efficient way of storing green crops in the 20–30 per cent dry-matter range, and so are the least dependent on having a prolonged spell of fine weather. Each crop can then be harvested as near as possible to the optimum date indicated by the D-value curve, and wilted in the field for 24 hours with the aim of harvesting at a minimum dry-matter content of 20 per cent, and a target of 25 per cent. At this target level of wilting the time that the crop is at risk to bad weather in the field is minimised; there is a useful reduction in the weight of crop that has to be handled; there is a good prospect of a stable silage fermentation; and there should be only a limited loss of effluent from the silo.

Chop-length

In general the shorter the forage is chopped in the harvesting operation the more of it can be packed into the trailer and the more rapidly it will consolidate during the crucial period while the silo is being filled. Short, uniformly chopped crops also create fewer problems when the resulting silage is being fed out; they are, of course, essential for loading and unloading a tower silo. Chopping also has nutritional advantages, because animals can eat more of short-chopped than of 'long' silage made from the same crop, particularly when they are self-feeding from the silage face.

The degree of chopping required depends on several factors, but it is generally agreed that material of 25–50 mm in length is suitable for most requirements. However, higher dry-matter crops are more difficult to consolidate, and the chop-length must then be shorter so as to get the degree of consolidation needed to prevent heating and moulding in the silo (see Table 7.1). While the more rapid consolidation with short-chopped high dry-matter material does reduce the risk of overheating, efficient sealing of the silo is probably even more important than with longer-chop, unwilted material. On the other hand, if a wet crop is chopped too short it may pack down tightly in the silo, and lose effluent which can contain much of the 'sugar' in the crop needed to produce a stable fermentation.

The chop-length of forage has often been described in terms of the harvesting machine used – single-chop, double-chop, etc. However, with the range of designs now available, and with the more precise control of chop-length that they allow, this classification is no longer adequate. One method of expression is in terms of *median chop length*: in a crop sample with a median chop-length of 25 mm, half the sample by weight will be in pieces longer than 25 mm and half shorter than 25 mm. This provides a single-figure basis for classifying chop-length. In other cases it may be important to know the quantity of material of a particular chop-length; the proportion of material in pieces longer than, say, 100 mm may be critical in determining the handling and flow properties of silage from a tower silo. Thus chop-length should be described in terms appropriate to the particular crop and silage system being used.

Types of forage harvester

In the early stages of the mechanisation of silage-making the *flail harvester* was widely used because of its simplicity, low capital cost and flexibility in use. However, most of the forage crops that were cut were loaded directly into the transport trailer, without any field-wilting. With the recognition of the importance of wilting crops before they are ensiled, flail harvesters were steadily replaced by machines which allowed the operations of cutting and lifting to be carried out separately. Flail harvesters had the further disadvantage that they exerted very little 'chopping' action, the median chop-length of the harvested crop generally being above 150 mm, and with many pieces longer than 300 mm; the suction action of the flails also readily picked up soil which could damage subsequent silage fermentation.

As described on p. 99, this led to the development of the *flail mower*, which returns the cut crop to the field in windrows that allow the crop to be wilted before it is lifted into the transport trailer. Initially this loading was done with a flail harvester, but these machines were gradually replaced by the *double-chop forage harvester*, which has a flail mechanism that feeds the crop into a cutter head, rotating at up to 1,000 rpm, and carrying knives which chop the crop before it is delivered to the trailer. These machines can directly harvest the standing crop, but they have mainly been used to pick up windrows of crop that have already been cut and wilted. Operated in this way there is also less risk of soil contamination than when the flails are used for the initial cutting operation.

Double-chop harvesters still give only limited control of chop-length in the harvested crop, and with more importance being placed on the nutritional advantages of short chop-length they have been replaced on many farms by *precision-chop (metered chop) forage harvesters*. These machines pick up the mown and wilted crop from the windrow and feed it either into a cutting cylinder fitted with knives which cut the crop cleanly against a shear-plate (front cover plate and Plate 7.1), or into an array of blades on a heavy flywheel (Plate 7.2). In both types of machine the chop-length of the crop is controlled by altering the speed of rotation and the blade configuration. The chopped crop is conveyed into the transport trailer by the airflow from the cutting mechanism.

Precision-chop harvesters are more expensive and complex than the earlier machines, but in general this is compensated for by their very high potential throughput – though this must, of course, be matched by the pre-cutting, field treatment and haulage elements of the overall operation. The driveshaft and gearbox of these machines can transmit considerable power to the cutting mechanism, and this demands a powerful operating tractor. The throughput is maximised when they pick up a large, even

Plate 7.1 Cylinder precision-chop forage harvester (CLAAS (UK))

Plate 7.2 Flywheel precision-chop forage harvester (Kverneland Kidd Ltd)

windrow at a slow forward speed. Selecting the forward speed that best matches the crop pick-up rate to the potential harvester throughput can be critical and requires a close range of tractor gear ratios at low speed.

The short-chop (typical median length 35 mm) produced by a precision-chop harvester allows a large quantity of crop to be carried in each trailer-load, an important factor when long hauls are involved. The chopped material is also easier to handle with loading machinery at the silo than when longer material is ensiled; it also consolidates more rapidly, giving better control of air movement and heating. The silage produced is very suitable for self-feeding systems, and it will also flow more uniformly than longer-cut material in mechanised feeding systems.

Precision-chop machines are generally fitted with a pick-up attachment for lifting wilted crops, but this can lift stones with the crop, particularly when several windrows have been gathered into a single larger windrow on stony land. These can blunt the chopping mechanism, and care is needed in the selection and operation of tedding and raking equipment. Under the worst soil conditions it may be necessary to restrict the harvester input to a single width of the pre-cutting machine, and this is one of the factors that has led to the development of mowers with a wide cutting width (p. 95) which are able to cut and set up a large amount of crop in a single windrow. The chopping mechanism can also be damaged by 'tramp' metal, and most harvesters are now fitted with a metal-detecting unit which disengages the chopping mechanism when metal is picked up; most cylinder machines are also fitted with a number of individually mounted blades which 'fail safe' with less severe damage than with full-width blades (Plate 7.3).

Larger models of this type of harvester have been equipped with a separate diesel engine, which allows the forward speed of the tractor to be continuously adjusted to match changes in ground conditions and crop without affecting the power available to the harvester, so giving more consistent output. Many contractors now use self-propelled harvesters, with an output that can be matched by few farmer operations.

Precision-chop machines are, of course, essential when the crop is to be loaded into a tower silo (p. 156), since the filling and unloading machinery will not deal efficiently with longer-chopped material. There are also crops, in particular forage maize, which can only be satisfactorily handled by this type of machine. However, while there remain considerable advantages in short-chopped forage in both the making and the feeding of silage in bunker silos,

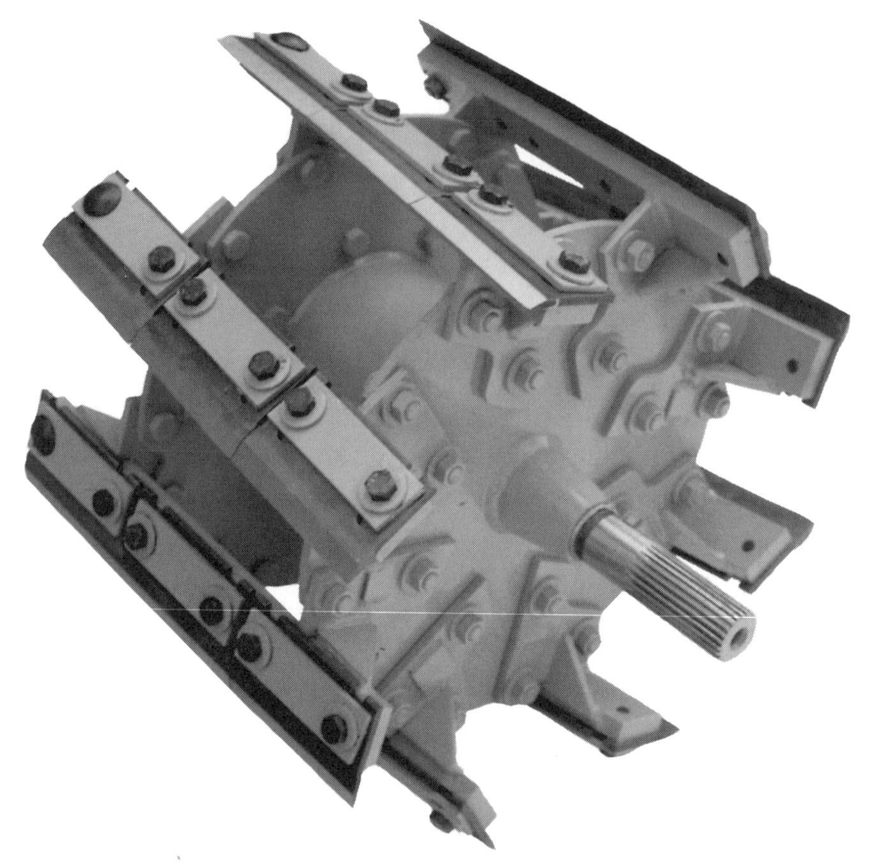

Plate 7.3 Chopping cylinder of precision-chop forage harvester
(Photo: CLAAS (UK))

in many situations a somewhat less precise specification of chop-length may be acceptable. This has allowed some simplification in design on some machines, with removal of the feed rolls fitted in front of the chopping mechanism. The effect of this change is most apparent when the harvester is set to give a very short chop-length (5 mm); the median chop-length can then be as high as 15 mm, compared with 10 mm when feed rolls are fitted, but there is much less difference at longer chop-lengths.

Harvesting forage maize for silage

The date of harvest of forage maize is somewhat less critical than with the grasses, because its digestibility remains fairly constant for several weeks around the optimum harvesting date. This stage, indicated by the grain becoming cheesy and firm, and a 'dent' appearing on the surface of the grain, is reached in southern England some time after the middle of September, depending on the variety and season, and somewhat later further north.

Whenever possible harvesting should be delayed until the dry-matter content of the crop has reached at least 25 per cent, and even then harvesting can be put off for a few days, if necessary, in order to wait for better weather; limited frost will help to dry the crop, but will not reduce its feeding value.

To ensure good fermentation in the silo forage maize must be finely chopped, because maize silage readily heats up and moulds if it contains large pieces which allow air to get into the mass. Flail and double-chop forage harvesters have had only limited success with forage maize, and the use of a precision-chop harvester is strongly advised; although some larger pieces will get through, the bulk of the crop will be finely chopped. The crop is now generally cut with a specially designed cutting head, which allows the height of cut to be adjusted (Plate 7.4). Many operators now aim to leave a 300–400 mm stubble because, although the weight of crop harvested is lower, this is likely to be more than

Plate 7.4 Forage maize head fitted to precision-chop harvester (Westmac Ltd)

compensated for by its higher digestibility, because it will contain less low-digestibility stem than a crop cut closer to the ground. Most maize harvesters are also fitted with a screen to crack the maize grain; this makes the grain more digestible, which is of increasing nutritional importance as crops are harvested at a more mature stage.

Most forage maize is now harvested with a precision-chop harvester fitted with a cutting head, noted above. Most 'farm' machines carry either a single or a double-row attachment; powered by a 70–100 hp tractor these have an output of 12–22 tonnes per hour, and can harvest up to 4 ha per day. However, an increasing number of maize-growers are now employing contractors to harvest the very heavy crops they regularly grow. Using self-propelled machines fitted with up to six cutting heads, and powered by 400 plus hp (Plate 7.5), contractors can harvest up to 20 ha of forage maize a day. It is then essential that transport to the silo, and loading the crop into the silo, match this rate of harvesting, and these operations are now often also carried out by the contractor.

Wholecrop cereals for silage

Wholecrop cereals can be cut at a fairly immature stage and ensiled by any of the methods for ensiling grass crops described

Plate 7.5 Self-propelled four-row maize harvester (John Deere Ltd)

below. At this stage of growth wholecrop generally contains enough soluble carbohydrates to give a stable acid fermentation without the use of a silage additive, but wherever possible the crop should be wilted in the field before it is ensiled so as to minimise effluent loss from the silo.

However, as described on p. 82, most wholecrop is now harvested at a fairly mature, higher dry-matter stage of growth, and preservation is by ammonia gas which is produced from urea mixed with the crop before it is loaded into the silo. The key to success, then, is the rapid sealing of the silo so as to retain the ammonia gas within the crop.

The preferred cereal crop is wheat, with its high grain to straw ratio, and long-strawed crops such as triticale should be cut so as to leave a fairly long stubble of lower-digestibility straw. The cereal crop should be direct-harvested at between 50 and 60 per cent dry-matter content, and short-chopping to about 20 mm length is essential to achieve adeqate consolidation in the silo. Thus harvesting with a precision-chop harvester is strongly advised. Urea is mixed with the crop, at 20–30 kg to each tonne of crop dry matter harvested. This is generally done by feeding the urea, either as pellets or in solution, directly into the delivery chute from the cutting mechanism on the harvester (Plate 7.6). It

Plate 7.6 Harvesting wholecrop wheat, with urea solution being metered into the delivery chute of the forage harvester (Gordon Newman)

may be possible to store wholecrop at even higher dry-matter levels if urease is mixed with the urea before it is applied.

The application of silage additives

There has been a big increase in the use of silage additives but, as discussed in Chapter 2, there is still considerable debate as to which type of additive to use, and when it should be applied. However, these decisions should be greatly helped by the information provided by the new Silage Additive Scheme (p. 26), while the guidance in Table 2.5 should also indicate the level of application needed for the particular crop being ensiled (in general, the wetter the crop and the lower its sugar content the more additive will be needed). Clearly the final decision will still rest with the individual silage-maker and, as we have indicated, the first priority must always be to wilt the crop before it is ensiled so as to reduce the need to apply an additive. If an additive containing a bacterial inoculant is being used it is also recommended that the viability of the product should have been independently confirmed by the Advisory Services; if the inoculant is being cultured on the farm a similar check will give confidence in the efficiency with which the process is being carried out. All inoculants must be maintained at low temperature and kept away from sunlight and air.

Once the decision has been taken to use an additive the key requirement then becomes a method of applying it uniformly to the cut crop. A number of different applicator systems are available, for both liquid and powder additives, including gravity feed mechanisms as well as positive feed pumps. It is probably wise to seek advice from the additive supplier as to the most suitable system for the product selected. Where only a few hundred tonnes of silage are to be made, equipment based on a 25 litre container feeding the chemical through a tube directly into the flail or chopping mechanism is satisfactory (as shown in Plate 2.1); it is then important to insert the pipe at a point where the harvester air suction will draw liquid from the feed nozzle and break it into droplets. Some manufacturers have failed to identify this position correctly, and adjustment may be needed to get the best location, while ensuring that the nozzle is not fixed at a point where back-pressure can cause spray and fumes to escape.

When larger amounts of silage are being made the additive is now generally carried in a 200 litre drum in a cradle fitted either to the harvester or to the towing tractor (Plate 7.7). Liquid is fed from

Plate 7.7 A 200 litre silage additive container with delivery tube to chopping mechanism (CLAAS (UK))

this into the cutting mechanism by a peristaltic or glandless pump, controlled either by the driver from the cab, or by an automatic cut-out device on the harvester pick-up. It is important to ensure that the pump can supply the additive at the required rate; thus a feed rate of at least 3 litres of liquid per minute is needed to apply 3 litres of additive per tonne to a crop being harvested at 60 tonnes per hour. The rate of application can be controlled either by the size of the nozzle at the end of the delivery pipe, or by varying the speed of the pump. In all cases the rate of flow should be checked against the rate at which the harvester being used fills the trailers, if possible by weighing typical trailer loads.

Great care is needed when acid (corrosive) additives are being used, and operators must follow the manufacturer's instructions, particularly with regard to the wearing of gloves and goggles during filling and changing containers. It is also important to ensure that the flow of additive is cut off whenever crop is not passing through the harvester (for example on headlands); many operators do not apply additive to the last few loads of the day so as to clear residual additive from the inside surfaces of the harvester – easier than the ritual washing down that is often advised.

The most important requirement with biological additives is to ensure that fresh and active bacterial cultures are used, and are renewed at least daily.

Accurate and uniform application to the crop is the key to the effective use of an additive. Thus there is real advantage in explaining to the harvester operators why the additive is being used – and, if possible, how it works. They are then much more likely to apply the additive at the correct rate, and to change the rate if the crop condition changes, for example with a sudden shower of rain. It is also very useful, as noted above, for a number of trailer-loads of crop to be weighed, for few farmers know how much their trailers hold – yet without this information deciding the correct rate of additive to apply can be very much a question of guesswork.

The silage produced with the use of an effective additive should be preserved better, and with lower losses, than silage made from the same crop but without an additive; it should also generally have a higher feeding value (see Chapter 9). In particular, animals can eat more silage when an additive has been used; thus more silage may be needed for the winter and, although the additive itself may have reduced losses during the making process, every other measure must also be taken to ensure that enough silage is made to satisfy this higher intake.

Forage harvesting systems

The output of a forage harvesting system is frequently limited by the available transport facility, and the importance of this part of the overall system cannot be overemphasised. Much depends on detailed aspects of preparation and planning which, though apparently obvious, are easily overlooked. Although, as has been noted, crop chop-length has a bearing on trailer capacity, other 'local' factors can also have a major influence, including the crop dry-matter content, the distance from field to silo and, above all, the skill of the operator.

Most forage harvesters can load either into a trailer towed directly behind the tractor or the harvester (see front cover and Plate 7.2) or into a towed trailer running alongside (as seen in Plate 7.1). In general the output per man-hour is higher with side-loading because it avoids the delay of coupling and uncoupling trailers (although this can be speeded up by using a hydraulic pick-up hitch), and this system is now the most commonly used so as to exploit the high output potential of the latest harvesters. In either case, when short-chop wilted crops are being loaded care must be taken to screen the top of the trailer to avoid loss of light and leafy particles. The delivery chute must also be adjusted to prevent losses when the harvester is turning corners, and in order to fill the trailer evenly from front to back the flap on the chute must be moved up and down; this is now generally done by electrical or hydraulic controls which change the angle of the flap.

A further factor in exploiting the high potential of these harvesters is the management of the cutting, tedding and raking operations in the field so that the crop, after it is wilted, is set up in windrows, slightly narrower in width than the pick-up on the harvester, and holding enough crop to match the throughput of the chopping mechanism. Different field routines will be needed, depending on whether the crop is kept mainly in windrows during the wilting process, or is spread over the whole surface of the field. In the former case two or more windrows can be raked into one, or a mowing routine such as that shown in Figure 7.1 can be used, with the mower-conditioner delivering alternate bouts to right and left into a single large windrow. In the latter case the spread crop, once it has wilted to the required level, must be raked into large windrows ready to be harvested (see Plate 5.6).

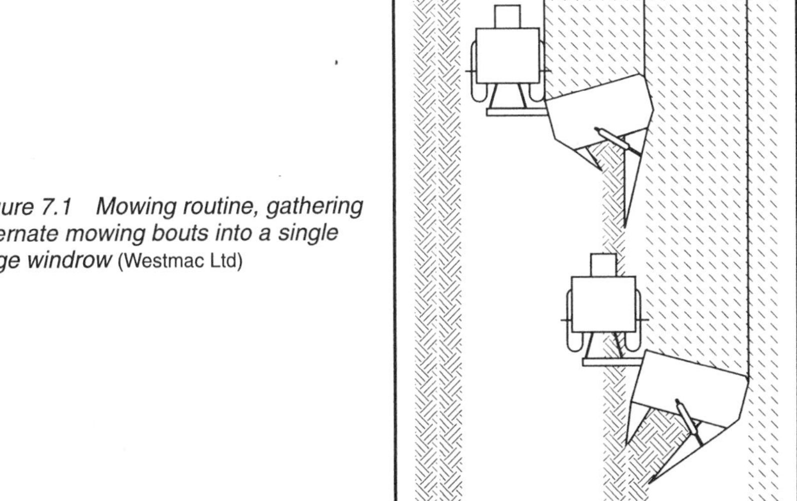

Figure 7.1 Mowing routine, gathering alternate mowing bouts into a single large windrow (Westmac Ltd)

Even when the organisation of field transport has been well managed, a too-common source of delay arises in the emptying of trailers at the silo, with badly designed trailer doors and catches jamming, so that force is needed to open the doors. The most effective type of door is the up-and-over counterbalanced gate, secured by a spring latch at the base; certainly the ease of opening under load should be one of the main considerations when a new trailer is being purchased. But even when the doors are open the crop may be so compacted that it can only be emptied at the silo by jerking the trailer forward on the tractor clutch (we must have all seen this many times!). To avoid this there is great advantage in the sides of the trailer being wider apart at the rear than at the front: standard trailers can be modified by fitting a smooth lining, widening towards the rear, within the fixed sides. When the trailer is tipped the load, even if compacted, slips out easily between these flared walls, and the time saved can more than compensate for the small reduction in the weight of crop carried.

Self-loading forage wagons

In its simplest form this equipment consists of a trailer fitted with a reel to pick up cut crop from the field and moving arms which pack the crop into the trailer. It is still used widely on small farms in southern Germany, where it is easier to operate on sloping ground than conventional harvesting equipment. This method of harvesting has been progressively improved, in particular by the addition of a chopper unit of 20 or more knives fitted between the pick-up and the packing arms, which greatly increases the trailer capacity, and also ensures much better fermentation when the crop is ensiled (Plate 7.8). Only a small number of these machines have been used on UK farms, in part because, despite their relatively low throughput (up to 15 tonnes per hour with a 500 m haul), a second man is generally required at the silo. However, a more recent design (Plate 7.9), with a precision-chop forage harvester directly coupled to a tipping trailer, has a higher output possibly more suited to farm size in the UK.

Avoiding soil contamination

It is impossible to overemphasise the importance of preventing the harvested forage being contaminated with soil; under extreme conditions forage may contain more than 20 per cent of soil, leading to a highly unstable silage fermentation. Soil may also contain listeria organisms, which can multiply and produce silage which can cause disease in livestock, most seriously in in-lamb ewes.

To prevent this all fields must be heavily rolled in the spring, and

Plate 7.8 Self-loading forage wagons are suited for harvesting crops for silage when little labour is available and transport distance is short

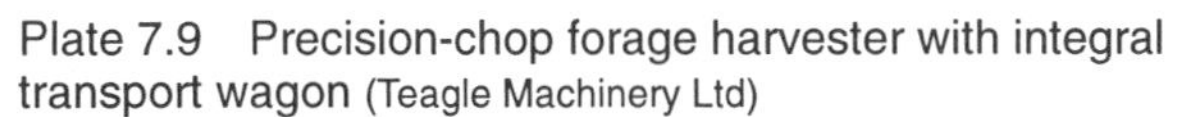

Plate 7.9 Precision-chop forage harvester with integral transport wagon (Teagle Machinery Ltd)

molehills levelled. The flail cutters on flail harvesting machines must be set so as to avoid skimming the soil, and the cutting heads on disc and drum mowers correctly sharpened and adjusted. Care is also needed to ensure that stones and soil are not picked up by the operation of the tedding, raking and windrowing equipment. Silage can also be contaminated from soil on the tyres of tractors loading the cut crop into the silo. The provision of a concrete loading area with a 1.2 m retaining wall at one side, against which the tractor buckrake can push, is a well worthwhile investment – and with outdoor silos this concrete surface will markedly aid both self-feeding and tractor loading of the silage in winter.

FILLING THE SILO

Most of the silage now made in the UK is stored in walled bunker silos, either indoors under cover, or in the open. In the past the construction of many bunker silos has been fairly primitive, and there have been frequent cases of walls collapsing under the high sideways pressure set up by a 2–3 m depth of wet grass loaded into the silo. Health and Safety Regulations now require a much higher standard of construction, both in design and materials, and including effective arrangements for drainage and the collection of effluent. Design is the business of professionals, and we do not attempt to deal with it here. But the farmer will have key inputs in specifying the dimensions of the silo (thus the width required for self-feeding is likely to be greater than for mechanical unloading), access, location of drainage, etc. His main task, though, once the silo is built, will be to operate a system of filling and sealing that will regularly produce a high-quality feed with the minimum of losses.

The Dorset Wedge system

The subjects of crop specification, stage of growth at cutting, and wilting and optimum dry-matter content have perhaps so far been considered rather abstractly. But at last the day will come when harvesting must begin. Simple matters, such as the overhaul of harvesters, the preparation of trailers and a check on the unloading gear may seem obvious, but if they have not been done there will be quite unnecessary delay. It is remarkable how often farmers leave silo preparation, and the purchase of polythene sheeting, until the very morning that cutting is due to start. In contrast, with all the equipment and the silo fully prepared during the previous weeks the whole operation of cutting and harvesting can swing into action on the date intended.

Indoor walled silos

As we have noted, most silage is made in bunker silos. These are now generally constructed with concrete walls, which must be inspected regularly and any signs of corrosion or cracking immediately made good, preferably with epoxy-based sealants. Long observation has shown that most of the wastage in walled silos occurs at the 'shoulders' where the silage, as it consolidates, sinks and pulls away from the the wall, and so allows air to get in and start decomposing the silage. This can be prevented by one of the key elements in the Dorset Wedge system, in which a length of 1.8 m wide plastic is sealed with mastic along the full length of the top inside face of each wall before filling starts (300-gauge polythene sheet is preferred, as thicker sheets are too stiff and heavy and tend to peel away from the wall). This sheet is folded back over the wall during filling (Plate 7.10). Then, as soon as filling is completed, the sheet is pulled across the surface of the silage, the top sheet placed in position to overlap it with mastic between the two sheets, and a top covering placed in position in order to hold the two sheets of plastic together (Figure 7.2). As the silage sinks the top of the side sheet peels away from the wall, but the weight of silage still presses the bottom half of this sheet against the wall, and so prevents the entry of air.

Unfortunately, long observation shows that not everyone follows this procedure, however desirable it may be, and silos with

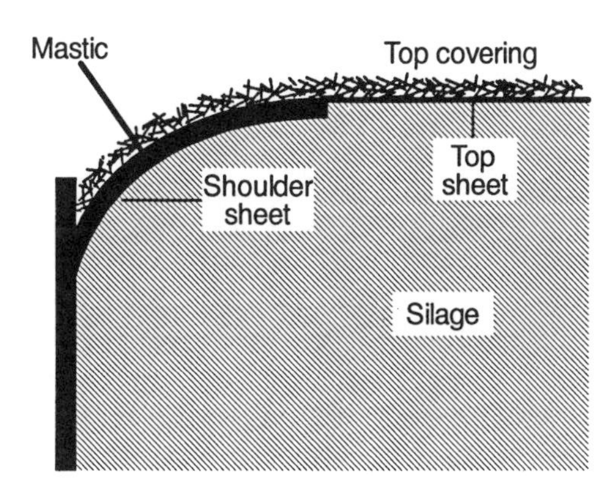

Figure 7.2 The use of side and top sheets to prevent wastage on the 'shoulders' of a bunker silo

Plate 7.10 Polythene sheet is sealed to the side wall with mastic and then folded back over the wall until filling is completed

'shoulders' blackened with wet, rotting silage are still a common sight. Some operators have used long cylindrical weights placed on the top sheet to hold it against the wall, but we have yet to see a generally effective alternative to the procedure shown in Figure 7.2.

The filling procedure

To minimise access of air, leading to oxidation and heating of the crop, the aim is to concentrate the loading of the crop into one section of the silo at a time, so that filling and sealing of that section are completed as quickly as possible. Thus the first loads that are brought in are dumped directly against the back wall of the silo, and packed close against each other, aiming for a depth of at least one metre against the wall on the first day. The following day at least a further one metre should be added so as to give consolidation and to act as a 'blanket' on the first day's fill. The need to ensure this minimum fill will determine the area covered each day, so in a silo of small dimensions it may be possible to fill to more than one metre over the whole floor area on the first day. In most cases, however, only part of the floor will be covered, and the silo is then loaded in the form of a wedge against the back wall, with a sloping face up which the loading tractor runs. The first day's fill will thus have the shape shown in Figure 7.3A. The slope established will depend on the width of the silo and the speed at which the crop is coming into store. Whenever possible a slope of about 20° should be aimed for, but slow filling of the silo may lead to a rather steeper slope – though with the key requirement that it must still be shallow enough for the tractor to run up and down with complete safety.

As soon as the last load is spread in position each evening the whole of the exposed surface of the crop must immediately be

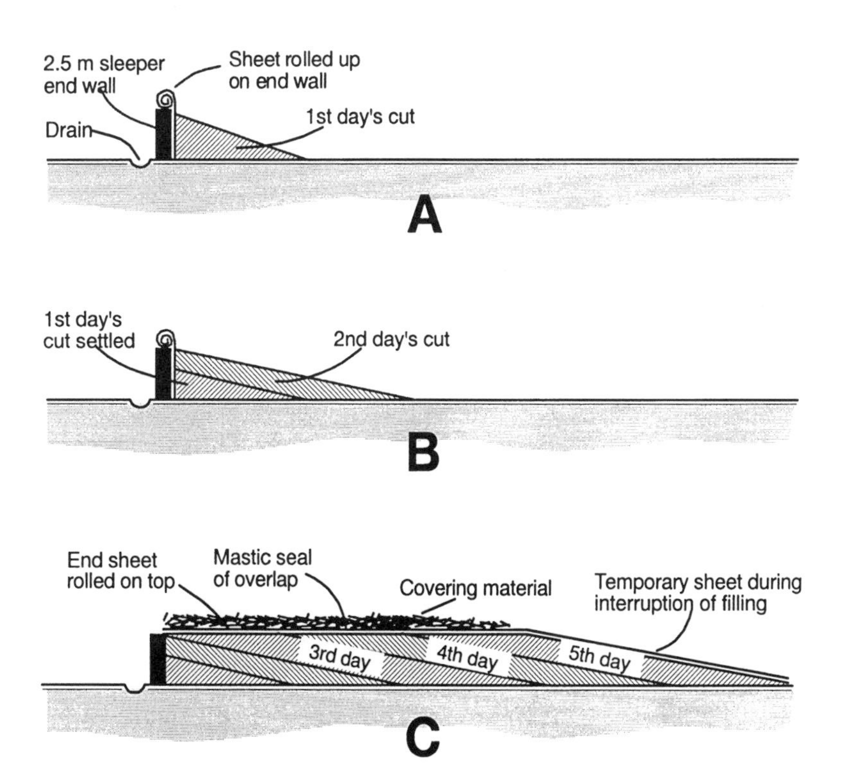

Figure 7.3 A to C: stages in filling a bunker silo by the Dorset Wedge system

covered with polythene sheeting; however, at this stage it is not necessary to pull the side sheets into position. *This top sheet does stop air getting into the crop; but its main purpose is to prevent hot air escaping from the crop and sucking in fresh air.* This is the most effective way (a) of stopping the sugars in the cut crop being lost by oxidation and (b) of creating the anaerobic conditions needed by the lactobacilli to ferment the sugars to lactic acid.

The top sheet is left in position overnight, and removed just before loading starts again the next day. Filling progresses as in Figures 7.3B and 7.3C, with a layer of not less than one metre of crop being added each day, while at the same time maintaining the original slope and progressing towards the open end of the silo.

The passage of the buckrake tractor as it loads fresh crop into the silo will give some degree of consolidation. But if the crop has been chopped to the correct length in relation to its dry-matter content (Table 7.1) the weight of the crop will quickly give a considerable degree of settlement, which then speeds up as the crop starts to ferment and the acid produced makes the plant cells collapse. There may be advantage in some rolling during filling, particularly if the crop is rather dry. However, as soon as loading finishes for the day, covering the surface of the crop with plastic sheet will generally be more effective than further consolidation in preventing access of air and controlling heating; in fact, further passage of the tractor over the surface may create a 'bellows' effect, as already noted, and draw in fresh air.

As soon as each section of the silo has been filled to the required height, generally level with or slightly above the side walls, the surface should be levelled off to give a uniform finish. The appropriate length of the two side sheets is then pulled over the 'shoulders' of the mass of crop, mastic is applied, and the top sheet laid in position and pressed down to give an airtight seal (Plate 7.11). Some weighty material must then be applied over the whole area of the top sheet so as to press it into contact with the surface of the crop and to expel as much air as possible. A 100 mm layer of chopped grass is very effective for this purpose, and is recommended if straw bales are to be loaded on top of the silage later, since the grass will prevent the sharp ends of the straw puncturing the sheet.

Levelling and covering progresses as the sections of the silo are filled, with particular care needed in sealing the final 'open' face of the silo. The aim is to ensure that any one section of the silo is filled from floor to finished height, and then sealed, *within three days*.

Plate 7.11 The top sheet is laid on top of the side sheets and sealed with mastic

This method has the particular advantage of allowing flexibility in organisation; in particular it should not be thought that the standards suggested demand a large labour force and lots of machines. In practice the Dorset Wedge system of filling has been applied successfully on many farms where one man has carried out the whole operation of cutting and filling. Equally it has proved effective in the hands of large operators, including contractors. The key to success is in being able to vary the rate at which the sloping face moves forward, in relation to the harvesting rate and the width of the silo, while keeping the silage fermentation under control under a wide range of conditions. There is also the great advantage that it is possible to discontinue filling the silo for a day or so if weather conditions make harvesting 'impossible', or over a weekend to allow staff a break, without risk of excessive heating.

Equipment for filling the silo

The cut crop is generally loaded into the silo by a tractor and buckrake; push-off buckrakes, either front or rear-mounted, are preferred. Rear-mounted machines have the advantage that they can more easily drive up the slope of the silo, but the driver does have to look backwards, which can be a strain. On larger operations a high-lift front-mounted buckrake with a working rate of up to 40 tonnes of crop an hour is most effective. However, while these machines allow accurate positioning of the loads of grass, there may be less immediate consolidation than when a tractor buckrake is used; particular care is also needed to avoid too steep a front slope, which can be unstable, and in covering the surface of the loosely packed crop each night so as to prevent movement of air.

Special-purpose handling vehicles, now operating on many farms, are well suited to loading silos, with their large-capacity forks, swift-acting hydraulics, and considerable manoeuvrability. Four-wheel drive is a distinct advantage if full output is to be realised (Plate 7.12).

Plate 7.12 Four-wheel drive vehicle for silage loading (Barclays Bank plc)

Outdoor bunker silos

It is a reasonable assumption that, if a satisfactory job can be done in protecting silage indoors from the damaging effects of air movement, then a similar method should also give protection from the weather in an outdoor unroofed silo. The main difference is that it is advisable to use a stronger, 500-gauge polythene sheet, or butyl rubber sheeting, to give the extra protection needed outdoors.

Much the same filling and covering procedure is used for outdoor walled silos as for indoor ones, but there is special need to ensure a weatherproof finish, and in particular to prevent the rain, which inevitably runs down inside the side walls, from penetrating into the made silage. Figure 7.4 shows an effective way of

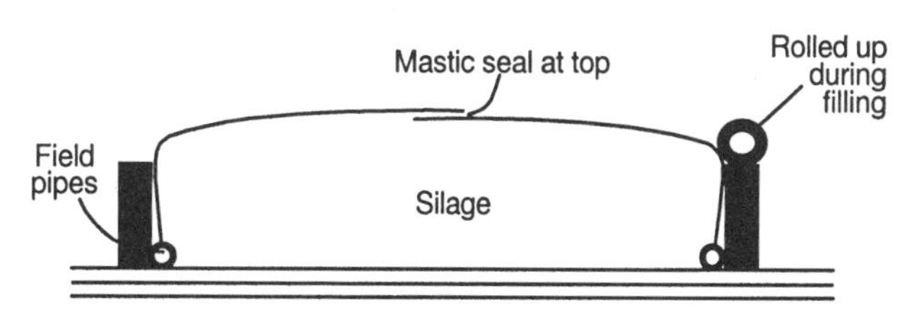

Figure 7.4 Sealing an outdoor walled silo to prevent the entry of air and rainwater

using polythene sheet for this purpose. Before filling starts sheeting is draped over each wall down to ground level, with the main 'cover' rolled up on top of the wall. The floor of the silo must slope towards the open end, and a field drainage pipe is laid the full length of each wall with the outlets taken to a trap for collection of rainwater and effluent (p. 160). Crop is loaded into the silo in a sequence similar to that shown in Figure 7.3, but ensuring that the final surface is slightly higher at the middle than against the walls. The surface of the crop must be covered with plastic whenever filling is not in progress; then, as soon as filling of a section is completed, the side sheets are drawn across the surface of the silage to overlap at this highest point, mastic is applied between the sheets, and the surface is then weighted down to hold the sheets in position. Again, old motor tyres are ideal for this purpose (Plate 7.13). The key to the success of this system is that rain falling on the silo should run down *between* the walls and the plastic sheets to the drainage pipes, and so cannot contaminate the silage.

Plate 7.13 Outdoor silage bunker well sealed with old motor tyres

A simpler form of silo can be made by constructing dwarf earth walls down the sides and along one end of a concrete pad; it is often possible to take advantage of local topography by excavating into the slope of a bank. Sealing is effected by lining the walls with polythene sheet down to the concrete, and overlapping this with a plastic top sheet, weighted in position. However, this system allows less control of the run-off of effluent and rainfall than with walled silos, and its use may be restricted in future.

The ultimate in economy and flexibility for storage is the outdoor unwalled silo, since this can be sited at any convenient point on the farm where there is a level site on clear ground. If the silage is to be self-fed the following winter a concrete base will be essential, and this could well be mandatory in future. Loading an unwalled clamp silo follows exactly the same principles as those described for an indoor silo, though with some simple adaptation to compensate for the absence of walls, and to ensure that the covering sheet is pressed firmly against the silage so as to prevent wastage. The silo sides should be formed with a slope no steeper than 20°. The position after the first day's loading shows how both the end slope and the filling slope are established (Figure 7.5A).

The second and subsequent days' fillings (Figure 7.5B) proceed with the aim of adding a minimum of one metre extra depth of crop each day, at the same time maintaining the shallow slope on the end wall and on the sides. As soon as enough length has been filled to the required height, and preferably after no more than two days, a 500-gauge polythene sheet is applied laterally across the silo (Plate 7.14) and immediately weighted down. As noted, old tyres are very effective for this purpose, but other covering materials, including soil and FYM, can be used (with FYM the final application can be made using a rotary muck-spreader driven round the silo). Whichever method is used, the whole surface of the sheet must be covered so as to prevent sunlight weakening the plastic, which is then sensitive to damage by birds. If soil or FYM is used the layer should be at least 150 mm deep to allow for the washing effect of rainfall; this is a further reason for the shallow slope of the walls. The covering material can also be more firmly anchored by making a light sowing of grass or cereal seeds, which will root and bind the whole together against washing by rain. It can also assist removal of the covering, since the whole mass will be rooted and can be pulled off in a mat.

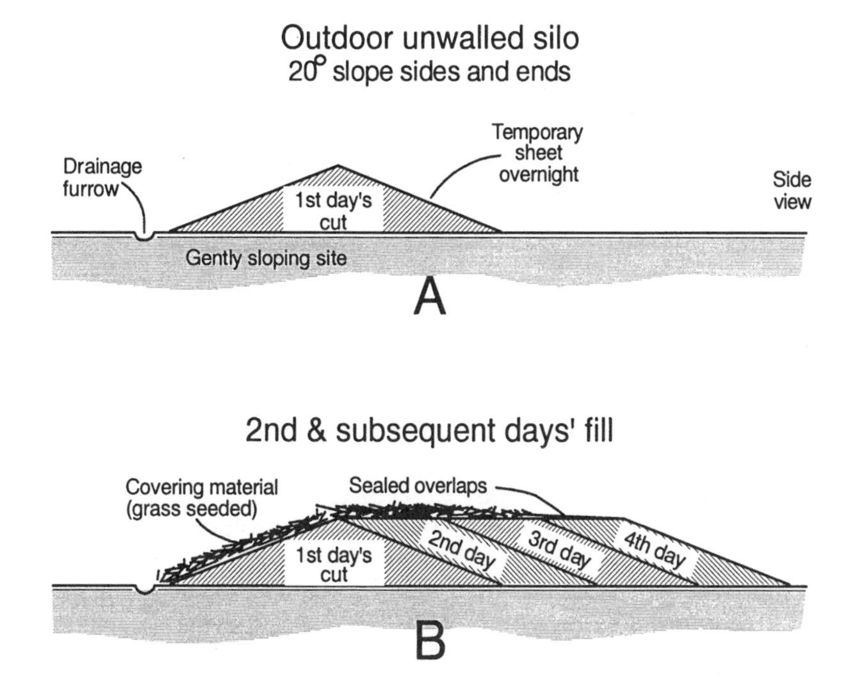

Figure 7.5 A and B: stages in filling and covering an outdoor sealed wedge silo

Plate 7.14 A 500-gauge polythene sheet being placed in position across the end of an outdoor wedge clamp silo immediately after the last load has been put in place

Prompt sheeting and weighting of the cover will go a long way to preventing surface wastage, even on the sloping sides where there will have been only limited consolidation. But any delay in sheeting will lead to heating and oxidation of the sugars in the crop; even if an additive is used there is then a risk of poor fermentation which will render the whole operation a waste of time. However, if future pollution regulations require that this type of silo must have the same effluent-collection facilities as walled bunker silos, this may detract from its apparent advantages of flexibility and cheapness; in any case it has now been largely superseded by big-bale silage.

Silage-making in big bales

Much of the experimental work on silage in the 1960s was carried out in sealed reinforced plastic bags holding about one tonne of fresh crop. When these bags were closed they were completely airtight, and the complete absence of waste in the silage when they were opened was a major stimulus to the development of the sealed bunker systems already described – and to the study of the practical possibilities of ensiling baled grass in plastic bags.

The first attempts were made with standard rectangular bales, but this was not adopted in practice because the bales required laborious manhandling, and losses were high because the bales did not fit tightly in the bags. However, the position was transformed with the introduction of large round balers (p. 115), producing bales of the order of 1.2 m width and up to 1.8 m diameter, and holding over half a tonne of green crop. These machines were initially developed for handling hay and straw, but with only minor modifications they have proved very effective for baling both wilted and unwilted green forages. They were initially used to bale 'long' forages, but many balers are now fitted with a pre-chopping unit, generally installed between the pick-up reel and the bale chamber (Plate 7.15). One model, fitted to the front of the tractor, lifts and chops the cut crop and forms a windrow which is picked up by the trailed baler.

Pre-chopping gives considerable benefits: the bales hold up to 25 per cent more forage than bales made from unchopped forage, so that baling and plastic costs per tonne of silage are lower; chopping releases the sugars in the crop, leading to a more rapid silage fermentation; and animals can eat more of chopped than of long silage – a particular advantage when the silage is being fed to sheep. Silage additives can also be more uniformly applied if they are sprayed into a pre-cutting unit than when sprayed on to the long crop as it enters the bale chamber.

Probably the major development, however, has been in the method of plastic sealing the round bales. Originally each bale was lifted by spikes on a tractor foreloader, and inserted into a heavy

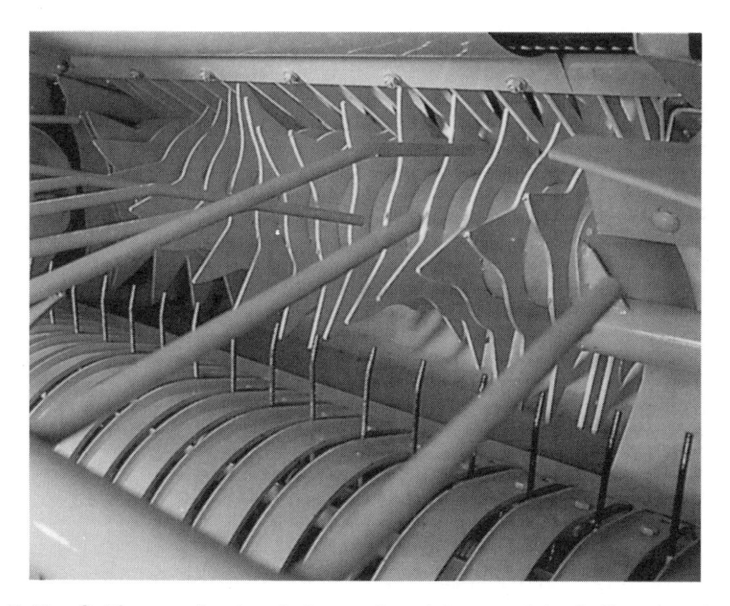

Plate 7.15 Cutting mechanism between the pick-up and the baling chamber on a large round baler (CLAAS (UK))

gauge round plastic bag which was placed in position on a stack of already-filled bags, so as to expel as much air as possible, before the neck of the bag was tied by hand. This was a laborious procedure; more seriously, many bags were pierced by the spikes, so allowing air to leak into the silage, causing considerable surface wastage.

In the last edition of this book, published in 1986, we described the first machines designed for the automatic wrapping of big bales in stretchable cling-film plastic; these have developed into a family of ingenious machines for wrapping big bales. Some are static machines, located alongside the silage store, and wrapping bales brought in from the field (Plate 7.16), so that the sealed bales are handled only once before they are stored. Others wrap bales picked up from the rows of bales left on the field behind the baler (Plate 7.17); yet others are towed directly behind the baler, with automatic control of the wrapping process (Plate 7.18). All use the same basic principle, however, of a series of rollers which rotate the bale in one direction so that it pulls a sheet of multi-ply plastic film under tension from a reel of film, while other rollers rotate the bale more slowly at right angles so that the film is wound round the bale in a spiral – with the speed of this rotation determining how many layers of plastic are applied before the bale is completely covered (generally the aim is to have at least four layers). The plastic is then cut and the loose end tucked in. Subsequent handling of the wrapped bales is best done with a side-gripper loader; if a spike loader has to be used any puncture holes must be sealed before the bales are stored.

Because the stretch film is applied under tension it contracts and maintains close contact as the silage shrinks during the storage period. As a result there is very little surface moulding and wastage, and recent research at IGER suggests that overall losses may, if anything, be lower than with even a well-managed sealed bunker silo system.

All the other basic principles of silage-making also apply. Whenever possible the crop should be rapidly wilted before it is baled. A minimum dry-matter content of 25 per cent should be aimed for, but crops up to 40 per cent dry-matter content can be ensiled in sealed bales with very low losses, while some operators have successfully stored what is in effect nearly-made hay at over 70 per cent dry matter. Thus cutting with a mower-conditioner is a great advantage, with subsequent tedding organised so that the final windrow is slightly narrower than the baler pick-up.

However, if weather conditions do not permit effective wilting

Plate 7.16 Static bale-wrapping machine (Parmiter and Sons Ltd)

Below:
Plate 7.17 Towed big-bale wrapper (Volac International Ltd)

Plate 7.18 Automatic towed round baler and bale wrapper (Rustons Engineering Co Ltd)

it may be better to bale a somewhat wet crop than to delay in the hope of wilting, for less effluent is lost from sealed round bales than from the same crop ensiled in a bunker silo, both because bales are generally stacked no more than three layers high (so that the overall pressure within the bales is less than in a silo), and also because the plastic wrapping on the bales retains much of the effluent that is produced. This is released, in relatively small amounts at a time, when the wrapped bales are opened for feeding.

As noted above, pre-chopping has many advantages, in particular in increasing the weight of crop contained in each bale, from the earlier 400–600 kg range to between 600 and 800 kg. Both fixed and variable chamber balers are used for baling green crops; as with hay (p. 116), the variable chamber machines produce more densely packed and heavier bales, but the greater cost and complexity of these machines perhaps make them more suited to contractors and to very large farms than to the general run of farms. On the other hand the 'swiss-roll' structure of bales produced in a fixed-chamber machine may have an advantage in making them easier to 'unwrap' for feeding out to stock.

Baled silage, if it has been efficiently wrapped, can be safely stored out of doors, preferably on a base of washed sand or gravel in a sheltered spot at least 10 m away from any ditch or watercourse. Some operators do store bales several layers high, but in general two or three layers are preferable so as to reduce the strain on the plastic wrapping (back cover). The stack should be baited against rats, and fenced to keep out inquisitive stock (Plate 7.19). It is also advisable to have an alternative source of feed available in case the bales freeze at low temperature.

The new generation of machines producing big rectangular bales (p. 118) has also proved rugged enough to bale green forages. The initial attempts to plastic-wrap these rectangular bales with machines designed for big round bales were not very successful, and where the bales have been ensiled this has generally been by covering a stack of bales with two sheets of plastic, held separately in place to the ground (Plate 7.20). However, a number of machines designed to wrap big rectangular bales have now been developed (Plate 7.21).

The use of these bales, which hold up to one tonne of forage, thus opens up a further option in large-scale silage-making. Interestingly the bales shown in Plate 7.21 were wrapped in white plastic film, while most round bales are wrapped in black film. Yet when black plastic is exposed to sunlight, either in the field or in a stack, it can become too hot to touch and, being under

Plate 7.19 Fencing is essential to keep stock away from big bales of silage (Volac International Ltd)

Plate 7.20 Big rectangular bales of grass ensiled under two separate layers of plastic (New Holland UK Ltd)

Plate 7.21 Automatic plastic film wrapper for large rectangular bales (Photo: Kverneland Kidd UK Ltd)

tension, may begin to become permeable to air – undesirable in a silage system. Perhaps white plastic should be used for all bale wrapping?

Clearly the potential of silage stored in big bales has been transformed from the situation we reported in the last edition in 1986 – 'that big-bale silage had proved useful in conserving smaller lots of silage on farms which had already conserved large amounts of silage (in bunker silos), so avoiding the need to open up an already sealed silo – as well as in introducing silage to farmers who did not consider their operation large enough for investing in conventional harvesting equipment'. Surveys in 1994 indicated that between 6 and 9 million tonnes of silage – representing about a quarter of total silage production – had been stored in sealed big bales, and the indications are that even more was made in 1995. This suggests that further developments in the mechanisation of ensilage in big bales will continue as one of the major growth areas in forage conservation.

Tower silos

We have paid particular attention to the ensiling of crops in surface walled silos and clamps and in big bales, because we believe that these will continue to be the principal methods used for making silage in the UK. However, in earlier editions of this book we examined in some detail the potential of tower silos – vertical cylinders into which the finely chopped crop is loaded, and in which the anaerobic conditions needed for silage fermentation are quickly established because only a small surface area of crop is exposed to air (p. 18). A few concrete towers were built just after the Second World War, but it was not until the mid-1960s that fully mechanised systems of livestock feeding from tower silos were introduced into the UK, based on the wide experience with tower storage of maize and lucerne silage that had been gained in the USA. Two main types of tower were installed – made either of reinforced concrete or of vitreous-lined steel – and both, when skilfully operated, gave very efficient preservation with low losses. Interestingly, much of this development was localised in certain areas, including northern England and the south-west coastal area of Scotland, where the enthusiasm of the first successful operators had encouraged others to follow.

A good deal of partisan controversy developed between the supporters of bunker silage and of tower silage. Our view remains that these do not represent different concepts of silage-making. Indeed, the more one examines the principles and practice of making tower silage, the more evident are the similarities with the standards recommended for bunker silage – in terms of stage of crop growth at harvest, wilting before storage, short chop-length, rapid filling, exclusion of air and prevention of wastage during feeding out. These requirements were already accepted as essential to the efficient operation of tower silos when they were introduced in the 1960s; as a result, this system immediately produced silages of good feed value with low losses. In contrast, most of the bunker and clamp silage being fed at that time was of low feeding value, and had been made with high losses.

However, the position changed with the introduction of the sealed bunker silage systems described in this chapter. In fact, filling a sealed bunker silo is in many ways similar to filling a 'horizontal' tower silo, with the main difference being that the crop for the tower silo *must* be more heavily wilted (to above 35 per cent dry-matter content) so as to reduce the pressure on the silo walls, and more finely chopped, so as to allow efficient operation of the filling and unloading equipment, than is necessary with bunker silos. This makes tower silo systems more weather-dependent than bunker silos.

Although few detailed measurements were made, practical experience of the two systems indicated little difference in the 'system' losses between well-operated sealed bunker silos and tower silos. Thus it became difficult to justify the higher initial capital cost of the tower silo, with its ancillary harvesting, loading and unloading equipment. As a result the installation of new tower silos effectively ceased by the late 1970s. However, the tower silos already installed have continued to operate very successfully and they are particularly well suited to the storage of both forage maize and wholecrop cereals. Recent developments in field wilting, and in new filling and unloading equipment, have also improved the reliability of tower silo operations, and towers are well suited to fully mechanised systems of feeding, including mixer-feeder wagons (p. 185). Thus we could see some new future interest in tower silos.

Ensiling forage maize

The growing of maize for silage has been transformed by the introduction of the new hybrid varieties, better adapted to the cooler regions of northern Europe; as a result, forage maize can now regularly be harvested in the southern half of the UK at dry-matter contents above 30 per cent, and at between 25 and 30 per cent dry-matter content further north. At this stage both the yield and the digestibility of the crop have levelled off at close to maximum values, so that the exact date of harvest is less critical than with grass crops.

Equipment for harvesting forage maize was described on p. 131. To ensure a satisfactory fermentation in the silo it is essential that forage maize is finely chopped, for the crop readily spoils by heating if it contains large pieces which allow air to get into the mass. Double-chop and flail-type harvesters do not give a small enough particle size and the use of a precision-chop harvester is strongly advised, preferably fitted with a screen (corn-cracker) to ensure that all the maize grains are broken.

However, the generally higher dry-matter levels at which forage maize is now being harvested appear to have exacerbated the problem of heating in maize silage, both in the making and storage periods and also during feeding out, as a result of fresh air getting into the silage as the face is loosened. Unfortunately none of the additives yet tested has shown any significant benefit in controlling heating in maize silage. (Of the seven bacterial and three enzyme maize additives in the UKASTA list, only limited laboratory data have been submitted on just four products.) This means that control of heating must depend mainly on management – on short-chopping the harvested crop; rapid filling of the silo, with the final height being reached in two days, rather than the three or more days indicated in Figure 7.3; efficient sealing, in particular at the shoulders; and minimal loosening of the silage face during feeding out.

Practical experience has shown that the risk of heating during both making and feeding is reduced if the maize is stored in a long narrow silo, which can often be achieved by installing a dividing wall in a conventional bunker silo. There may also be some reduction in heating during feeding out if the exposed face is sprayed daily with propionic acid. And to make a more general point: the control of heating becomes more difficult the higher the dry-matter content at which the forage maize is harvested, yet there is little evidence of any nutritional benefit from ensiling at above 35 per cent dry matter, and no effluent problem at this level. This suggests that the crop should be harvested in the 30–35 per cent range, and that waiting for higher dry-matter levels to be reached may create unnecessary problems.

Ensiling wholecrop cereals

As was noted on p. 34, as soon as urea is mixed with the chopped cereal crop ammonia gas begins to be produced. It is then essential to load the crop as fast as possible into a silo fitted with side sheets, as in Figure 7.2, to roll to give some consolidation, and then to seal to prevent the ammonia gas escaping. Because ammonia is heavier than air particular care must also be taken in sealing round the bottom edges of the stack, and in the final sealing of the whole surface area. The space outside the silo must also be well ventilated so that any ammonia that does escape cannot affect people working on the silo. At least three weeks, and preferably longer, should elapse before the silage is fed, to give the ammonia full time to react with the crop. As noted above, tower silos are well adapted to storing wholecrop cereal, and give very efficient retention of the ammonia gas produced when urea is mixed with the crop.

Effluent loss from silage

When forage is ensiled the sap in the plant cells is rapidly released, and then provides the main source of sugars and other nutrients for the fermenting micro-organisms. However, if the crop is very wet, and particularly if it is heavily consolidated, some of this moisture can be squeezed out of the heap of forage in the form of silage effluent. More than 100 litres of effluent (over 10 per cent of the fresh weight of crop) can be lost as effluent from a tonne of crop loaded into the silo at 20 per cent dry-matter content, and losses as high as 200 litres per tonne have been measured with even wetter crops. In contrast, very little effluent is lost from crops that have been wilted to above 25 per cent dry matter before they are ensiled (at 20 per cent dry-matter content juice can readily be squeezed out of the cut crop by hand; at 25 per cent very little juice is released).

The dry matter lost in the effluent is generally less than 1 per cent of the dry matter in the crop (small in relation to some of the other losses in the overall ensiling process), but it is a valuable fraction, containing readily digestible soluble nutrients and minerals. However, the main problems are that silage effluent causes the social nuisance of smell and, more seriously, the risk of pollution of water-courses. The huge increase in silage production during the past two decades has coincided with – and some would say has been partly responsible for – the growing public awareness of the dangers of river pollution, and with it the intensive monitoring of water quality by the National Rivers Authority (now the responsibility of the new Environmental Agency). This monitoring, coupled with targeted advisory programmes as in the Vale of Torridge, has been an important factor in reducing the incidence of water pollution by silage effluent and by animal slurry. Cases of pollution do still occur, however, and at high legal cost to those responsible.

Thus the prevention of effluent contamination of water-courses must now be a top priority on livestock farms. This means first producing as little effluent as possible – and then ensuring that any effluent that *is* produced does no damage. Quite the most effective way of reducing effluent production is to wilt the crop before it is ensiled, with a target of 25 per cent dry matter, at which relatively small amounts of effluent are produced unless the silage has been finely chopped or stored at a depth greater than, say, 3 metres. As we have recognised, if adverse weather conditions prevent adequate wilting it may sometimes be preferable to load a crop into the silo at below 25 per cent dry matter rather than leaving it in the field to deteriorate. While effluent production will be unavoidable, the aim must then be to minimise the quantity of effluent by avoiding heavy consolidation of the silage, and by ensuring that the silo is completely sealed – for heating and the oxidation of sugars produces water which can further increase the moisture content of the crop.

Low-moisture crops will generally have been ensiled with an additive, but there is very little experimental evidence on the effects of additives on effluent flow. (Only two of the 106 products listed in the UKASTA Approval Scheme (p. 26) claim any benefit under Category C3, Reducing Effluent.) There is thus likely to be some effluent flow from most silo operations at some time during the year; the imperative then is to ensure that this effluent cannot cause pollution.

With indoor silos the effluent is intercepted, as it flows from the base of the silage face, by a half-round drain, laid across the fall of the floor at the lowest end of the silo, and draining into a catchpit. The vertical walls of the silo should be water-tight (and inspecting them and sealing any leaks should be an essential spring task), but in case there are any leaks a drain should also be laid alongside the outside base of each wall (as in Figure 7.3).

Outdoor silos present a bigger problem because, as well as the effluent from the silage, the drainage must also be able to deal with rain that falls on the plastic sheets covering the silo. As shown in Figure 7.4, these sheets should also be in place between the silage and the walls, so that rain that runs off the top sheets will be channelled to the base of the wall and thence by perforated cylindrical drainpipes to the main outflow to the catchpit. All drains outside the silo must be securely covered to prevent them becoming blocked with loose silage or slurry.

Getting all the effluent into the catchpit is a considerable achievement. But it is not enough, because the catchpit may not be large enough to hold all the effluent produced from a single silo, and it is most unlikely to hold a whole season's output. It must thus be emptied at intervals – preferably fairly frequently, so as to minimise the risk of overflowing, and also because fresh effluent is more pleasant to handle than the evil-smelling mature product. Most effluent is disposed of by distributing it back on to the fields from which the forage was cut. However, there have been reports that this can cause scorching. Recent work in Northern Ireland suggests that this may be most marked on swards that have been allowed to regrow before the effluent is applied – leading to the recommendation that effluent, at up to 50,000 litres per hectare, can be applied soon after the sward is cut with little risk of scorching (and with some apparent advantage from the N, P and K returned to the soil). The effluent is best applied with a dribble-bar from a slurry tanker; it must not be sprayed because of the risk of nuisance from smell.

Alternatively, some of the fresh effluent that is produced can be fed to livestock – though there are unlikely to be enough animals to deal with it all. For this the effluent is pumped daily from the catchpit into a 2,000 litre plastic drinking trough, sited at a point where the animals have ready access. Under practical conditions dairy cows have consumed up to 5 litres of effluent a day – with presumably some nutritional benefit. However, the effluent must be fresh, and any surplus must be disposed of daily.

Care during storage of silage

Six months or more may elapse between sealing the silo after it is filled and opening it for feeding out, and during this time it must be inspected at intervals to ensure that the seal remains intact, and that the covering sheet is kept in close contact with the silage underneath – for even if the silo appears to be well 'sealed' there will always be wastage wherever there is a gap between plastic and silage. Outdoor clamps should always be fenced off to prevent them being walked on by cattle or sheep – or even deer – with disastrous results to the sheet, and to the silage below.

The silo should also be checked to see that no rain is getting in, and that there is no faulty drainage of the site which can lead to waterlogging of the lower layers of silage. Straw bales are often stored on top of indoor bunker silos and care must be taken when loading these to avoid damaging the covering sheet. At the time the silo is sealed it is also useful to mark with coloured sacking a number of points on the surface where the covering sheets overlap, which allows later use of a core sampler to take silage samples for analysis, without damaging the sheets.

Considerable care is needed when a retaining wall has to be removed to expose the silage for self-feeding or mechanical unloading, because of the pressure the silage may be exerting on the wall. Hopefully, as the wall is removed the polythene lining sheet should be revealed still firmly adhering to the silage inside, which should be completely free from surface wastage. The days when the farmer had to spend the best part of a day cutting out and hauling away rotted material before edible silage was revealed must surely be a thing of the past. The management of the silage face so as to prevent wastage during feeding out is discussed on p. 175.

Developments in the technique of big-bale silage over the last ten years (p. 152), in particular the replacement of plastic bags by cling-film, have greatly improved the efficiency of the system and reduced the wastage that often occurred as a result of the silage shrinking away from the walls of the bag. However, the plastic film remains susceptible to physical damage during handling, and the precautions we have noted are strongly advised.

Silage with restricted fermentation

It was noted in Chapter 2 that a few silage additives act by partly sterilising the crop, rather than by just relying on a reduction in pH. A number of these additives contain formalin, and the resulting silages differ in both appearance and smell from most acid silages. The procedure for making the silage is identical to the methods already described, with rapid filling of the silo in daily increments followed by prompt covering and sealing with polythene sheet. However, because of their low acid content, these silages are more vulnerable during long-term storage to damage from spoilage organisms and from oxidation, and particular care is needed to prevent air getting into the silo, and to minimise the free surface area when the silage is being fed out. A noticeable feature of these silages is the absence of the typical silage 'smell' – possibly of some relevance with the greater emphasis now put on environmental impacts. There is also some evidence that less effluent is produced as a result of the reduced fermentation during the ensilage process.

Some of the methods we have described for making bunker and clamp silage, and the precautions when making big-bale silage, may seem elementary – even obvious. But their aim is to produce, every year, a predictable feed of high quality, and with low losses. One has only to make a most cursory check on many current silage-making operations to see that, because these precautions are not being taken, serious surface wastage still occurs – and with it the losses associated with overheating and effluent flow from the silo. This represents one of the most serious wastes of expended resources in the whole range of farming activities – crops that have been grown and harvested, at some expense, and then wasted. Yet in many cases only quite minor changes in methods and timing can yield impressive results in reducing losses and improving the quality of the silage that is made. This is within the grasp of every silage-maker.

CHAPTER 8

STRAW AS ANIMAL FEED

Cereal straw has always been used in animal feeding, but its low crude protein content, poor digestibility – the D-value of straw is generally below 50 – and low intake by ruminants has limited its usefulness in rations for high-producing animals. Straw is indigestible because the fibre it contains is firmly bound together by lignin – which gives the straw the strength it needs to hold up the heavy ears of ripening grain. When the straw is eaten this lignin protects the cellulose and hemicellulose in the fibre from attack by the fibre-digesting micro-organisms in the rumen. It has been known for many years that when straw is treated with alkali its digestibility is increased, because the alkali dissolves away some of the lignin from the ligno-cellulose complex and at the same time makes the fibres swell. Both these factors make the fibre more accessible to digestion by the rumen organisms.

Interestingly, much of the initial work with alkali treatment of straw, as with silage additives, was carried out in Scandinavia, although the main method used there, in which the straw was soaked in sodium hydroxide (caustic soda) solution and excess alkali washed out before feeding, was applied on only a relatively limited scale. However, the high animal feed prices of the 1970s led to renewed interest in the possible role of treated straw in livestock feeding in the UK, and several novel processes, using both sodium hydroxide and ammonia, were developed. These were more effectively mechanised, and so required less hand labour, than the original process; mechanisation also meant that more concentrated alkali, and in some cases heat, could be used, making the delignification process faster and more complete. In the case of sodium hydroxide it was also found that it was not necessary to wash excess alkali out of the treated straw, because in practice this caused no harm to the livestock being fed, and could in fact be of nutritional advantage (p. 200).

This renewed interest in straw treatment coincided with much more straw becoming available as a result of the huge increase in cereal production which followed UK accession to the European Union. Thus by the early 1980s annual straw production was well over 12 million tonnes – at least 7 million tonnes more than was required for conventional bedding and feeding, potato storage, etc. Yet, despite this ready availability of straw, only a small fraction has been used for animal feeding. There were two main reasons for this. First, it has generally been more convenient to burn the straw in the field (a much easier process than baling and carting, and also giving good control of cereal weeds and diseases), and after the straw-burning ban in 1990 it has still continued to be easier to chop the straw behind the combine and plough it in. Second, most of the surplus straw is produced in the south and east of the country, while most of the livestock that could use it are in the west and north – and straw is costly to transport over long distances.

Yet plenty of cereal straw is in fact available within reasonable distance of many livestock farms, and more straw is now being baled and fed – with a spectacular increase during the drought in 1995. In particular some larger dairy farms are now using mixer wagons (p. 183) which make the feeding both of chopped straw and of alkali-treated straw products more practical (and in which the residual alkali in the treated straw can help to buffer the acid in the silage included in the feed mixture).

The digestibility of straw

Straw can differ in digestibility, with the straw from spring barley and oats in general being more digestible than wheat straw, and straw from spring cereals being more digestible than that from winter cereals (possibly because stronger straw is needed to support the heavier winter crops). The application of a cereal fungicide prolongs photosynthesis in the flag leaf, and may increase the digestibility of the straw, as will leaving a long stubble at harvest; conversely, weathering in the field is likely to make straw less digestible. The digestibility of straw can be increased by treatment with alkali, to levels similar to that of medium-quality hay or ensiled mature grass; under experimental conditions treatment with sodium hydroxide is more effective than with ammonia (Table 8.1).

Collecting and storing straw

The aim of the many investigations carried out since the early

Table 8.1 The digestibility (D-value) and crude protein content of cereal straw before and after treatment with either sodium hydroxide or ammonia

	D-value	*Crude protein (%)*
Barley straw	42.9	4.2
+ sodium hydroxide	61.8	4.5
+ ammonia	51.5	7.7
Wheat straw	41.1	3.9
+ sodium hydroxide	53.4	3.6
+ ammonia	49.1	6.8
Oat straw	48.2	3.4
+ sodium hydroxide	64.1	3.2
+ ammonia	55.3	7.5

(Data: ADAS, 1994)

1970s has been to develop practical methods of treating straw consistently and economically. Straw is a bulky, low-value commodity which must be collected and delivered at low cost if it is to compete with alternative feeds. Clearly it is cheapest when it is used on the farm where it is grown, for transport can increase its real cost two or three times. Standard rectangular and round bales are acceptable over short distances, and may be conveniently handled by farm transport at intervals over the winter. However, over longer distances high-density big bales, weighing more than 200 kg/m^3, are preferred. These can double the load of the average lorry, and many more loads of big straw bales can now be seen moving westwards, with much of the straw being baled by contractors so as to achieve the 2,000 tonnes plus annual output needed to make the use of high-density balers economical.

Straw intended for alkali treatment must be clean; it should contain less than 20 per cent of moisture; and it must be protected from rain, both before and after treatment. Where ammonia is to be used (see below) there is advantage in treating the straw soon after harvest, and the plastic sealing needed to retain the ammonia gas within the stack of straw should give effective protection until the treated straw is fed.

Feeding untreated straw

Most of the straw currently used is untreated, often fed directly from big bales either on the ground or in racks. Particularly for growing young stock and in-lamb ewes it can be usefully supplemented with maize gluten or a source of protein and energy such as molassed urea feed blocks or licks. In the case of sheep, 'browsing' the straw, though apparently wasteful, allows the animals to select a more digestible feed by rejecting the coarser stemmy fraction. Limited amounts of straw can also be included in 'complete' diets for dairy and beef cattle (p. 184), in particular to stretch out a dwindling silage supply in late winter. As most mixer wagons cannot cope with long straw this straw must be chopped; however, chopped straw is a very low-density product, both for storage and for handling into the mixer wagon, and this is a further reason why much of the straw fed in complete diets is now treated before it is fed.

Alkali treatment of straw

Both ammonia and sodium hydroxide can be used as delignifying agents to increase the digestibility of straw. While sodium hydroxide is the most effective (Table 8.1), the majority of straw treatment to date has been with ammonia, which is applied either in the anhydrous form or as a 35 per cent aqueous solution. This is now generally done with individual large plastic-wrapped round or rectangular bales, with the ammonia injected at the rate of 30 kg per tonne of straw at 20 per cent moisture content (Plate 8.1). Alternatively a number of large unwrapped round bales can be inserted into a long plastic sleeve before the ammonia is injected (Plate 8.2). The holes through which the ammonia was injected must be carefully sealed and the bags checked at intervals during the storage period until the treated straw is fed (Plate 8.3). The rate of delignification depends greatly on the ambient temperature, the process taking up to four weeks in mild weather and six to eight weeks in winter.

Analysis by ADAS of numerous samples of ammonia-treated straw has shown an average D-value of 54, similar to that noted in Table 8.1. The table also shows that ammonia treatment increases the 'crude protein' content of straw, as a result of chemical reactions between the ammonia and constituents in the straw which produce non-protein nitrogen compounds; these provide a source of 'rumen-digestible protein' (see Figure 3.2) which may reduce the need for protein supplementation when the treated straw is fed.

It is estimated that some 100,000 tonnes of straw are treated annually with ammonia, mainly in sealed big bales. The bales

Plate 8.1 Injecting aqueous ammonia into a plastic-sealed large round bale of straw

Plate 8.2 Straw bales being loaded into a plastic 'tunnel' prior to injection of ammonia (Gordon Newman)

Plate 8.3 Sleeves containing large round straw bales injected with aqueous ammonia

should be opened a day before feeding to allow excess ammonia to escape. The treated straw remains stable and palatable to stock (but not, apparently, to birds and rodents!) for three or four days; it has proved a particularly valuable feed for suckler cows, and for the rearing of 'framey' heifers with good rumen capacity.

There is also now more interest in alkali treatment with sodium hydroxide, both because this increases the D-value of straw more than does ammonia (Table 8.1), and also because the treatment is much faster. However, sodium hydroxide is a dangerous chemical, highly corrosive to the skin, and extreme care is needed when it is utilised, including the use of protective clothing and goggles. For this reason straw treatment is often carried out by contractors, using specialised equipment.

In the first process introduced in the UK hammer-milled straw was sprayed with a concentrated (27 per cent) solution of sodium hydroxide applied at the rate of 5 per cent of the dry weight of straw. However, this system used small bales and had a high labour requirement, and it has been largely replaced by bigger capacity machines which process large round or rectangular bales. Thus one machine (Plate 8.4) hammer-mills the straw and then meters an accurate application of sodium hydroxide solution which is rubbed into the straw with a special double auger; the treated straw is blown on to a heap which heats up for several days as delignification proceeds. This equipment, which can process up to 8 tonnes of straw per hour, is being operated by several contractors in the south-west of the UK. Perhaps surprisingly, it has not been used in the north-east of Scotland, despite the high population of dairy cows and the ready availability of wheat straw in that region.

In an alternative system, used mainly in the Midlands and east of England, the alkali is sprayed on to straw in a tub grinder, but this does not allow as accurate a control of the rate of application of the alkali. In all cases the treated straw must be stacked under cover for several days to allow delignification to be completed – indicated by the straw turning a golden colour with a plate-like structure, with no 'fluffy' material.

Feed mixer wagons are now used on many dairy farms for complete ration feeding (p. 183), and some of the later models, which are relatively water-tight and are fitted with external bearings, can also be used for alkali treatment of animal feeds. They were first used to mix whole cereal grain with 'pearl' sodium hydroxide, added at the rate of 30–50 kg per tonne of grain. After storing for four days the product, 'Sodagrain', has been

Plate 8.4 Large round straw bale being loaded into a 'Stropper' prior to treatment with sodium hydroxide (Gordon Newman)

found to be more completely digested by cattle, and to provide a better source of slowly degradable starch, than the equivalent grain treated by rolling or grinding. A Somerset farmer, Ian Ham, extended this process by adding 500 kg of chopped straw, together with further alkali and some water, to each tonne of treated grain. The mixed feed (Sodamix) is stacked for four days to allow maximum delignification, and promises to provide a useful new (alkaline) feed ingredient to be combined with low pH acid silage in complete diets for beef and dairy cows.

However, practical experience has shown that the high sodium content of these alkali-treated feeds does increase the water consumption of the animals being fed, and so the amount of urine they produce. This may not be a problem in cubicle housing, where slurry may be rather solid, but it can be a disadvantage with animals housed on deep litter, and there will then almost certainly be a need for additional bedding (and so for more straw). Care must also be taken to minimise the sodium content of the rest of the ration, and additional magnesium may need to be fed, because sodium reduces the availability of this essential mineral element.

In the context of the total forage supply, treated straw represents one of the most promising opportunities for more efficient feeding of ruminant livestock in the United Kingdom. Large amounts of the raw material are readily available, new techniques make field handling, transport and treatment more economic, and recent practical experience confirms the high feeding potential of the product. In particular, the flexibility of straw treatment in response to the problem of feed shortage in a dry summer, and in bridging the feed gap in a late winter, underline the potential importance of this new feed resource.

CHAPTER 9

METHODS OF FEEDING

The first aim in conserving forage is to store a sufficient quantity of feed, of appropriate digestibility and intake characteristics, for the stock that are to be fed. This operation must be carried out efficiently, with the minimum of losses, and in a way that takes as much account as possible of the later requirements for feeding out. The method of storage and feeding adopted will, of course, depend much on the layout of the farm and the numbers and type of stock to be be fed. But whichever method is used it is essential that it allows animals to eat the planned amount of feed, while at the same time minimising physical wastage and deterioration of the feed. Much of the skill and effort put into conserving forage can be wasted if the method of feeding is not carefully planned and carried out.

HAY AND STRAW

In the past much hay has been wasted during feeding because it has been considered to be – and often has been – of too low value to warrant much care. This has been particularly the case with outwintered stock, which need only a maintenance level of feeding, and bales of hay have just been spread across the field, with inevitable losses from treading and contamination from dung and urine. Even less care has been taken when straw has been fed to outwintered stock. However, with the increasing 'real' costs of both hay and straw, in particular on farms some distance from cereal-growing areas, more attention has been given to reducing this wastage, and more of these feeds are now being fed from racks or troughs. These do concentrate the effects of treading, which can be bad for both soil structure and the animals' feet, but these disadvantages are often accepted as a cheaper alternative than providing a hardcore standing or a temporary shelter for the stock, with the extra work and lack of flexibility these can introduce.

Much hay and straw is fed to stock in loose-housing, often in buildings that are outdated and with difficult tractor access, and under these conditions 'man-handleable' standard bales provide a very convenient method of feeding. These bales are generally fed from racks or troughs, and to reduce wastage from animals dropping feed these should be lined with a layer of weldmesh so as to restrict the amount of hay or straw that can be pulled out at one time. Sheep pose a particular problem as they tend to reject much of the hay they take, and quite a small gauge mesh is needed to minimise this selective feeding behaviour.

Much more hay and straw is now being fed that has been stored in large round bales. These can be handled by a tractor foreloader, either directly into cylindrical profile feed racks, or placed behind a feed barrier along a feed passage. However, stock can waste much of the feed that they pull out of intact bales. So, whenever practicable, the bales should be opened up before they are fed, either by unrolling them along the floor of the feed passage (Plate 9.1), or by using a machine that breaks up the bales and delivers the hay or straw into the feed troughs in chopped form (Plate 9.2). Fed in this way there should then be little restriction on feed intake and, with some tidying up by hand, feed wastage should be low.

The feeding of hay and straw from large rectangular bales requires a more open layout to allow the bales to be moved directly by tractor foreloader from store into a V-shaped rectangular cage feeder in the feeding area. Again this should have weldmesh sides, matching the dimensions of the bale, against which the bale presses as the hay is eaten. Alternatively these bales can be set down in the feed passage, where the separate wads of hay can be readily moved by fork in front of the animals.

Clearly it is not possible to specify the optimum method of feeding, because this will depend both on the type of stock to be fed, and on the building layout and access. However, the aim should always be to design a method which eases the labour requirement but which ensures that the stock are able to eat their planned ration without wasting feed.

SILAGE

Silage is, and is likely to continue to be, the main form of conserved forage made in the UK, and considerable effort is being given to developing ways of feeding silage that give high intake

Plate 9.1 Equipment for 'unwrapping' large round bales of hay or silage for feeding in the field or in feed passages (Parmiter & Sons Ltd)

Plate 9.2 Equipment for shredding and feeding out large bales of straw, hay or silage, indoors or in the field (Teagle Machinery Ltd)

with low wastage. Most silage is made in walled bunker silos, sited either indoors or outdoors, and the silage in these is either fed out directly from the open face in some system of self-feeding (Plate 9.3), or is mechanically extracted and moved for feeding at another site. Silage made in outdoor clamp silos is generally transported for feeding, as self-feeding is less practicable.

Plate 9.3 Dairy cows self-feeding at the silage face

Silage now makes up a major part of the winter ration on many livestock farms. This can involve feeding a thousand tonnes or more of silage over the winter months, and self-feeding is often preferred because it avoids the need to install, and operate, separate unloading equipment and feeding arrangements. Self-feeding systems do, however, require careful planning and operation, in particular to ensure that the stock are able to eat the required amount of feed. The first point to ensure, of course, is that intake is not limited by the silage being of low quality; but there are a number of other causes of low silage intake. The animals may find it difficult to pull silage away from a heavily compacted face, particularly if the silage has been made from long forage. This poses a particular problem with young animals and animals with poor teeth, hence the priority we have placed on the short chopping of forage before it is ensiled. Intake can also be restricted if there is not enough space at the silo face. Thus when stock have 24-hour access to the silage, a face width of 0.05 to 0.08 m is advised for in-lamb ewes, increasing to 0.15 m for young cattle and at least 0.2 m for adult cattle and dairy cows. However, if only limited access time is allowed, and particularly if all the stock have to feed at the face at the same time, the face width for adult cattle should be increased to at least 0.7 m.

As well as ensuring adequate feed intake the method of feeding must avoid physical wastage and must also minimise the risk of the silage deteriorating as a result of air penetrating the exposed face. An electrified wire or bar positioned across the face is the most common way of controlling the rate at which the silage is eaten back. The bar is moved frequently so that fresh silage is continually available, the aim being for the silage face to move back at least 0.15 m a day so as to prevent it heating and moulding. The risk of heating is much greater when high dry-matter silages, and in particular wholecrop cereal and maize silages, are being fed, and the silage face should then be fed back much faster, if necessary by restricting the width of the face that is fed. Railway sleepers or telegraph poles placed a couple of feet in front of the silage face help to keep silage behind the electrified bar and reduce contamination of the silage by animal droppings; slurry must, of course, be regularly scraped away from the area of the silage face.

Where an electric fencer is used to energise the bar or wire, care must be taken to ensure that the stock do not get too severe an electric shock. Some fencer units produce a very powerful shock, which can deter young and nervous animals and markedly reduce their silage intake. A power level that will jump a 3 mm gap is fully adequate, and if the output is much greater than this, and cannot be adjusted, an electrician should be consulted.

As noted, self-feeding is most effective with short-chopped forage. A 25–50 mm chop-length is generally adequate, and intake is higher if the silage is not too heavily consolidated. The silage should also be as uniform as possible, and care is needed to prevent wastage if the silo contains layers of silage made from different crops. In particular, any moulded patches on the surface or shoulders of the silage must be removed because, if moulded silage is mixed with the main bulk, it can greatly reduce the amount of silage animals are willing to eat – hence the priority given to complete sealing as soon as the silo is filled. It is also important to ensure that the face is fed back uniformly, because overhanging layers of silage make the face unstable, particularly in a nearly empty silo.

With modern equipment there has been a steady increase in the finished depth of silage in the silo, and in many cases this is now above 2 m, which is as high as even adult cattle can reach. It is then advisable to 'easy-feed' the top layer, by hand-forking it down behind the sleeper barrier. This is particularly necessary when sheep are being fed, and the feeding face should then not be higher than about 1.2 m.

Mechanised feeding

A number of different systems are now used for the mechanical extraction and feeding of silage from bunker and clamp silos. These offer the following advantages:

- They allow greater flexibility in the siting of silos, which need not be directly adjacent to where the animals are housed and fed.
- Silage can be stored to a greater settled height than is practical when self-feeding systems are used; silage unloaders are now able to operate with face heights of 3 m and more.
- The silage can be weighed and mixed with other feeds before it is fed.
- Each silo can be used for feeding several different groups of stock, which is seldom possible when the silage is being self-fed.
- Silage from two or more different silos can be mixed before feeding.
- The silage can be stored in unroofed bunker or clamp silos, which are cheaper than roofed silos but are seldom suited to self-feeding.

The mechanised feeding of silage is now being used for all classes of dairy and beef cattle, and it has also been successfully used for in-wintered sheep. Selection of the most appropriate system from the wide range now available must take account of a number of factors. Thus the method of operation and the capacity of the unloading equipment must be related to the height of the silage face, and matched to the number of stock, the daily amount of silage that is to be fed, and the distance from the silo to where the stock are housed. The type and dimensions of the feeding equipment will also depend on the building layout and access, the width of the feeding passages and the height of the feed troughs. Finally a forage box or mixer-feeder wagon is indicated if it is planned to mix the silage with other feeds before it is fed, or if the amount fed is to be metered. The following paragraphs note features of some of the systems and equipment now available.

Silo unloaders

Tractor foreloaders, which are simple and versatile, are often used to unload from the silage face. However, the tines do tend to open up the face of the remaining silage, letting in air which can cause aerobic deterioration and moulding. Thus hydraulic grabs are now commonly used, with a second set of tines which are forced down vertically through the silage and which allow a block of silage to be pulled away without too much disturbance of the face. Foreloaders can also transport silage from the silo to feed troughs with minimum spillage; alternatively they can be used for loading silage into a forage box or feed mixer.

Block cutters

These machines have a number of reciprocating knives which cut blocks of silage that can then be removed with little disturbance to the exposed silage face. Models mounted on the rear of a tractor can deal with a silage face 2.5 m high (Plate 9.4), while fore-mounted machines have a higher reach of up to 4 m. These machines cut blocks of silage ranging from 0.85 to 2.5 m^3, and weighing between 300 and 1,000 kg and can transport and drop the silage directly in front of the feed barrier. Some hand-forking may then be needed to break up the blocks and to ensure that all the silage is eaten, and some operators now use equipment designed for feeding baled silage (see below) for distributing silage from blocks. Silage blocks can also be fed using a forage box, but this is not generally advised as their solid weight puts considerable strain on the floor conveyor and beaters.

These machines have now largely been superseded by the hydraulic shear grab (Plate 9.5). This has horizontal tines which are pushed into the silage face, and a shearing blade which is then forced down vertically through the silage, cutting a block of silage which is removed as the loader reverses away from the face. With skilful use these machines leave a clean vertical face with no

Plate 9.4 Silage block cutters can handle all types of silage: they leave a compact and undisturbed face

loosening of the remaining silage (Plate 9.6); because of their shearing action they can also deal effectively with silage made from crops that have not been short chopped. Different models cut blocks ranging in weight from 450 to 1,000 kg. These blocks can be dropped directly on to a feeding passage, shredded and distributed into a feed passage or feed trough (Plates 9.2 and 9.7), or loaded into a forage box or mixer wagon, again with precautions to avoid overloading.

On many farms the total winter forage requirement is now harvested by contractors and stored in outdoor silos; a shear grab unloader may then be the only specialist machinery on the farm committed to silage-making and feeding.

Feeding big bales

As noted in Chapter 7, up to a quarter of the silage now made in

Plate 9.5 Hydraulically operated shear grab unloader for silage (Parmiter & Sons Ltd)

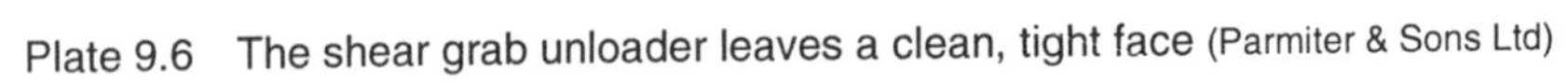

Plate 9.6 The shear grab unloader leaves a clean, tight face (Parmiter & Sons Ltd)

Plate 9.7 Equipment for shredding and delivery of straw, hay or silage from large round bales or from blocks of silage (Taarup Kidd Farm Machinery)

the UK is stored in big round bales. Originally the bales of forage were stored individually in heavy duty plastic bags, and to feed out the silage each bag had to be opened, the bale of silage lifted with a tractor-mounted spike and the bag removed by hand, before the bale was deposited in a feeding passage or feed trough. An advantage of this system was that, with care, the bags could be saved and used again the following season, but overall this proved a time-consuming and laborious operation, and in many cases the bags were just ripped open and thrown away.

The use of plastic bags has now been replaced by mechanical wrapping of the bales with cling-film plastic. Removing the plastic, and the original twine-binding from the bales does, of course, still involve hand work, but less care is needed than when bags were used because no attempt is made to reuse the plastic (though it should be recycled, together with all other farm waste plastic, through the Farm Film Producers Scheme: 0800 833749!). Various devices have been developed for unrolling the silage from the bale along the feed passage (as seen in Plate 9.1) and, with a limited amount of hand-forking, the silage can be fed out with low losses. When the plastic is removed from silage made in big rectangular bales the silage breaks up into 'wads' which can be easily hand fed (Plate 9.8).

If silage is to be fed from troughs, the bales can be distributed by a tractor-mounted machine, noted above, which shreds the silage and delivers it into the trough via a side conveyor (as in Plates 9.2 and 9.7). The same machines can be used to transport and deliver baled silage into racks in the field for outwintered stock, and they can also handle large round bales of hay and straw.

Plate 9.8 'Wads' of silage from large rectangular bales are easy to distribute by hand (New Holland UK Ltd)

Forage boxes

With larger herds feeding equipment with even greater capacity may be required, and forage boxes, ranging in size from 3–10 m^3, and holding up to 5 tonnes of silage, are commonly used (Plate 9.9). These are filled at the silo with a tractor-mounted foreloader or shear grab; a slat conveyor on the floor of the forage box then moves the silage forward into a set of rotating beaters; these tease out the material on to a cross-conveyor which delivers it to the feeding area. On some machines delivery can be to either side, which is of advantage if the stock are housed in buildings with restricted access. While forage boxes can deal with most types of silage they operate best with short-chopped material. If silage blocks are being loaded the blocks should be cut to a fairly small size so as not to put excessive strain on the conveyor.

Several different feeds can be fed at the same time by loading

Plate 9.9 Forage boxes, generally filled by a tractor foreloader, are well suited to feeding larger herds

them in layers into the forage box and using the unloading beaters to give some mixing as the feed is unloaded. However, mixing is only partial, and where complete mixing is required a mixer-feeder wagon is generally used. Some makes of forage box are also fitted with a hopper above the cross-conveyor which can deliver concentrates and other free-flowing materials on to the silage as it is being distributed into the feed troughs.

Mixer-feeder wagons

As the specifications of the optimum rations for different classes of livestock become more precise the livestock feeder may need to control the composition and quantity of feed that his animals eat more accurately than is possible with current self-feeding regimes. With very high-producing animals there may also be nutritional advantage in feeding the different ration ingredients together in a single mixture, rather than feeding forages and concentrates separately, as in most present systems.

Thus there is increasing interest in feeding systems in which the different feed components are weighed and thoroughly mixed together before they are fed, generally *ad libitum* so as to fully exploit their intake potential. As has been noted, partial mixing of feeds can be carried out with a forage box, but specially designed mixer-feeder wagons are needed to give complete mixing of a wide range of livestock feeds. These generally consist of a large open-topped metal tank, into which the different feeds are loaded, fitted with a device which thoroughly mixes the feeds before they are discharged by a side-conveyor. Mixing is generally done either by a set of counter-rotating augers in the base of the tank (Plate 9.10) or by a combination of augers and paddles (Plate 9.11). Some models also have a chopping unit which allows limited amounts of long silage, hay and straw to be included in the mixed ration. In order to avoid a heavy starting load the augers should be rotated slowly during the whole time the mixer wagon is being loaded. The mixed feed is then unloaded via a side-conveyor into troughs in the feeding area (Plate 9.12).

These machines produce a feed mix from which animals find it difficult to select out individual feed components (sheep are more adept at this than cattle), although when long forage or large pieces of fodder beet or concentrate cubes have been included some selection may occur. Many mixer-feeder wagons are also fitted with electronic weighing devices, generally based on load cells, and often with automatic recording, which allow the weights of the different feed ingredients in the mixed feed to be monitored and recorded. The same equipment can also be used to control the amount of a feed mix that is delivered to each group of animals; most importantly, it can produce and deliver different feed mixtures, each tailored to the specific nutrient requirements of a particular group of livestock.

Mixer wagons range in size from 4–12 m^3; the density of the mixed feed is typically of the order of 300 kg/m^3 (indicating loads weighing between 1 and 4 tonnes), but this can vary considerably with different feed mixtures. Most of the mixer wagons operating in the UK are towed units, powered by tractor pto (10–40 hp load), but truck-mounted units are widely used on feedlot operations in the USA, and some may already be operating on large livestock farms in the UK.

Feed passages and feed troughs

The design and dimensions of the arrangements for feeding silage

Plate 9.10 Contra-rotating augers in a mixer-feeder wagon (Taarup Kidd Farm Machinery)

Plate 9.11 Combination of augers and paddle in a mixer-feeder wagon (Opico Ltd)

need special care so that the amount of silage the animals can eat is not restricted, yet neither is feed spoiled or wasted. Two main methods are used: floor feeding behind a feed barrier, generally with yokes so as to prevent the cattle moving their heads sideways and so throwing feed about; and feed troughs, again often fitted with yokes. In the case of feed troughs, approximately 0.45 m per animal of trough length is needed for young cattle, increasing to 0.5 m for cattle at 18 months and up to 0.7 m for mature cattle. It may be possible to reduce these lengths when complete mixed rations are being fed, because competition between animals then appears to be less. The width of the trough should be related to the reach of the animals, with up to 0.8 m being suitable for dairy cows, with the base of the trough 0.3 m above floor level. The height of the front retaining wall will depend very much on the method of distributing the feed. It must be lower than the delivery chute on the forage box, feeder wagon or bale shredder (all generally above 0.6 m). Feed troughs should have smooth internal surfaces free from internal obstructions that can harbour moulding feed and with no uprights that can obstruct discharge from the feeder.

Plate 9.12 Side-delivery from mixer-feeder wagon (Taarup Kidd Farm Machinery)

Similar lengths per animal are required for floor feeding, but stock find it easier to throw feed forward than with troughs, and feed must be forked back against the retaining wall at intervals to prevent wastage. In all cases when high-yielding animals are being fed it is important to remove any uneaten silage at least daily; this can be fed to less demanding stock.

Mechanised feeding from tower silos

The key to efficient operation is the unloading equipment which cuts and extracts the stored silage. Both top and bottom unloaders are used, depending on the type of silo, and in the early days of tower silos the unreliability of this equipment presented the main operating problem, often made more difficult because the crop was not chopped sufficiently short before it was ensiled. However, current unloading equipment is much more reliable – although it is still advisable to have a stand-by electric generator in case of power cuts – for a full tower with a 'dead' loader promises some heavy hand work.

The silage from the unloader can be delivered into a forage box or mixer wagon at the base of the tower, or on to a conveyor system which distributes it to the livestock feeding area. The latter has several advantages:

- Feeding can effectively be fully automated and controlled by electronic weighing and timing devices, although precautions must still be taken to ensure the safety of both staff and livestock.
- There is no regular requirement for tractors and drivers.
- Building space can often be more effectively used, as there is no need for wide feeding passages and turning space, and headroom is less critical than when forage boxes are used.

However, despite these advantages, relatively few conveyor feeding systems have been installed with tower silos in the UK. This is partly because conveyors are fixed installations, and so less flexible than forage boxes, but also because they give only limited mixing when other feed components are added from hoppers on to the silage on the conveyor. Conveyor systems have also been proposed for feeding from bunker silos, but this involves loading into a forage box and thence on to the conveyor, with little apparent operating advantage.

A wide range of feeding systems and equipment is now available for feeding hay, straw and silage. Many of these give a high degree of mechanisation, allow other feed components to be mixed with the forage before it is fed, and can be equipped with automatic weighing and recording devices. However, while there have been important innovations during the last decade, probably the most important developments have been in improving the *reliability* of the equipment that is required to supply feed to livestock over long periods, often under appalling weather and ground conditions. This equipment demands rugged design and construction – coupled with regular and careful maintenance.

CHAPTER 10

FEEDING CONSERVED FORAGES

In Chapter 3 the main factors determining the nutritive value of conserved forages were examined, and it was shown that both hay and silage can have a high potential for animal production. However, it was concluded that productive livestock are most unlikely to be fed only on conserved forage, and that hay and silage will generally be fed in a mixed ration with other feed components. The key to successful feeding, then, depends on understanding, and exploiting, their nutritional interactions with the alternative feed resources, including combinations with other forages that are available.

Forage conservation will continue to be based on two main methods of preservation – storage of the crop in dried form as either field-made or barn-dried hay, plus small quantities of 'dried grass'; and storage of the undried crop under conditions either of low pH (silage) or of high pH (alkali treatment). Figure 10.1 summarises the results of many comparisons that have been made of the efficiency of recovery of dry matter and feed value when forage crops have been conserved as either hay or silage. These studies have shown higher field losses for hay (as would be expected, because crops cut for hay remain on the field for a longer time than crops for silage); higher in-store losses with silage than with hay (the result of the fermentation and effluent losses during the ensilage process); and then a further widening in terms of feed value resulting from the generally lower digestibility of crops that are cut for hay than for silage, as well as of the overall higher nutrient losses in haymaking, noted in Table 3.2.

Thus hay, even when made under 'good' conditions, yields only about 85 per cent as much feed value as silage. This lower efficiency of haymaking, coupled with the unreliability of weather conditions for making hay, has led to the present situation in which silage is now the principal method of forage conservation in the United Kingdom (see Figure 1.2). The following sections thus deal mainly with the feeding of silage.

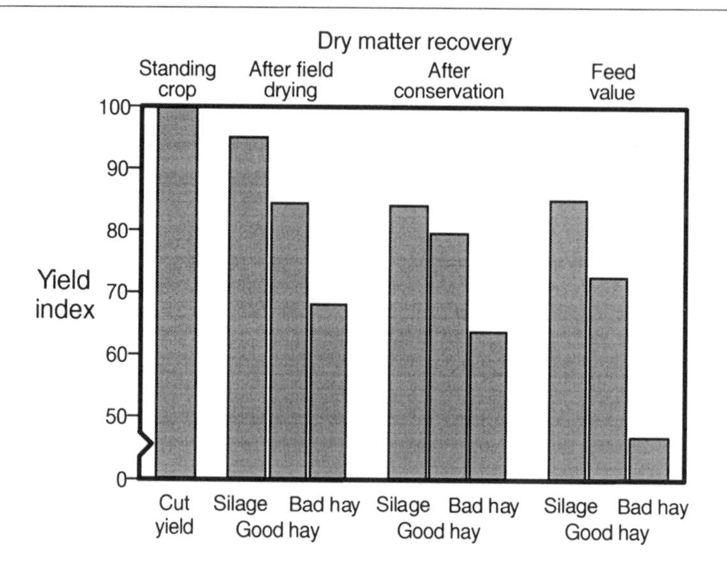

Figure 10.1 Typical relative yields of forage (dry-matter and feed value recovery) by different conservation methods (Data: Lingvall: Swedish University of Agricultural Sciences)

CONSERVED FORAGES IN DAIRY COW FEEDING

Figure 1.1 showed that the average milk yield per dairy cow in the UK rose at about 2 per cent per year up until 1972, but then at double that rate until milk quotas were imposed in 1984. Yet despite this marked increase in milk yield after 1972 the rate of concentrate feeding fell, from 0.4 kg per litre of milk produced in 1972 to about 0.33 kg per litre in 1984. This meant that forages were making up an increasing proportion of the rations that were being fed. One of the main reasons for this was that hay was being replaced by silage as the main conserved forage that was fed to dairy cows, the direct result of the wide adoption of the improved silage methods described in Chapter 2.

Not only was the silage that was fed of higher digestibility than the hay that it replaced, but it also made up a higher proportion of the total ration because in many cases dairy cows were able to eat more dry matter in the form of silage than of hay. This led to an increased requirement for conserved forage and, as a result, some 50 per cent more forage was being conserved in 1984 than had been conserved in 1972 (see Figure 1.2).

With the introduction of milk quotas in 1984 the new priority was to produce each litre of milk as cheaply as possible. Dairy farmers responded by further increasing their winter feeding of silage, and some 20 million more tonnes of silage were made in the UK in 1994 than in 1984.

One of the main changes that has resulted from the shift from hay to silage has been that crops are being cut and conserved at a less mature, more digestible stage than was previously practical with haymaking. Yet the full potential of earlier cutting, which in the case of *dry* conserved forage leads to an increase in both digestibility and voluntary intake (e.g. Table 3.5), may not always be realised when silage is fed, because of the often poor relationship between the digestibility and the intake of silage, as noted in Chapter 3. Many experiments have, of course, shown the benefit of early cutting for silage, leading to high levels of animal production. Thus in the research reported in Table 10.1, milk production was 3 kg per day higher from early-cut, high-digestibility silage than from silage made from the same crop but cut at a more mature stage, and so of lower digestibility, when both were supplemented with the same amount of concentrates. However, while in this experiment the intakes of the early-cut and late-cut silages were similar, in other cases there has been little or no advantage from the higher digestibility of the earlier-cut silage, because animals have eaten *less* of this silage than of the silage made from later-cut material.

This was shown most clearly in a series of experiments, carried

Table 10.1 Milk production from silage of high or low digestibility, fed with the same amount of concentrate supplement

	Silage, high digestibility	*Silage, low digestibility*
Silage D-value	74	64
Voluntary intake (kg DM/day)		
– silage	9.3	9.5
– concentrate	6.3	6.3
Milk yield (kg/day)	28.0	24.7
Milk fat (%)	3.61	4.10
Milk protein (%)	3.15	2.94

(Data: GRI)

out by IGER and supported by the old Milk Marketing Board, in which dairy cows were fed three silages, made respectively from an early-cut, high-digestibility crop ensiled efficiently with formic acid (HDGood: 75 D-value); from the same crop ensiled less efficiently without formic acid and without overnight sealing (HDPoor: 77 D-value); and from the same sward cut and ensiled three weeks later at a more mature, lower digestibility stage (LDGood; 64 D-value). Figure 10.2 shows that when the silages were fed to dairy cows as the sole feed the intake of the well-preserved silage (HDG) was no higher than that of the less digestible silage (LDG), but that both were more than double that of the poorly preserved silage (HDP). As a result daily milk production from the high-digestibility but low-intake silage HDP was the same as that from the less digestible silage (Figure 10.3; the interactions with levels of concentrate feeding are examined below).

Chemical analysis of the two high-digestibility silages showed that HDG contained 24.3 per cent dry matter, 19.0 per cent crude protein and 11.7 per cent of ammonia N in the total N, and that silage HDP had a lower dry-matter content (22.6 per cent), 19.6 per cent CP and a slightly higher ammonia N (13.1 per cent) content; these differences were quite inadequate to explain the two-to-one difference in voluntary intake between the two silages. It is discrepancies such as these which have given urgency to the detailed analytical studies on methods for predicting silage intake, reported on p. 50. These methods are being steadily improved, and are now claimed to be able to account for about 85 per cent of the variation in voluntary intake between grass silages.

However, no matter how accurate analytical methods are for predicting silage intake potential, as well as the D-value and Metabolisable Energy content of silages, they are 'being wise after the event'. They can help the livestock farmer to feed the silage that has already been made more effectively; but what is also needed is greater confidence, at the start of the silage-making operation, that the silage that is going to be made will be of the required D-value, and that when it is fed animals will eat a lot of it.

Information such as that in Chapter 4 should be a useful guide to the digestibility of the silage that will be made, but more reliable advice on how to ensure that the silage will have a high intake potential is still needed; present evidence points to the key importance of increasing the dry-matter content of the crop before it is ensiled, of getting a rapid fall in pH during the initial stages of the process so as to minimise protein breakdown, yet at the same time avoiding too high an acid content in the silage that is finally made.

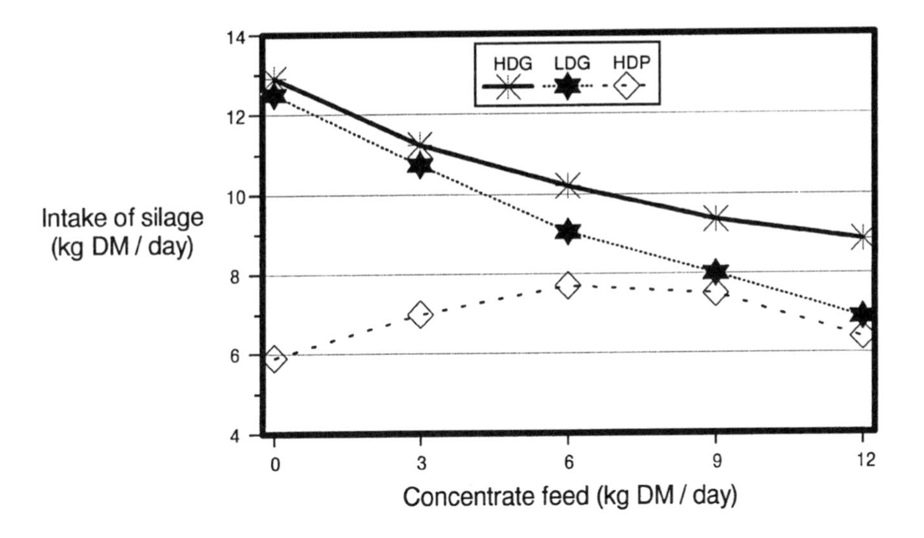

Figure 10.2 The dry-matter intake of three silages by dairy cows at different levels of concentrate feeding

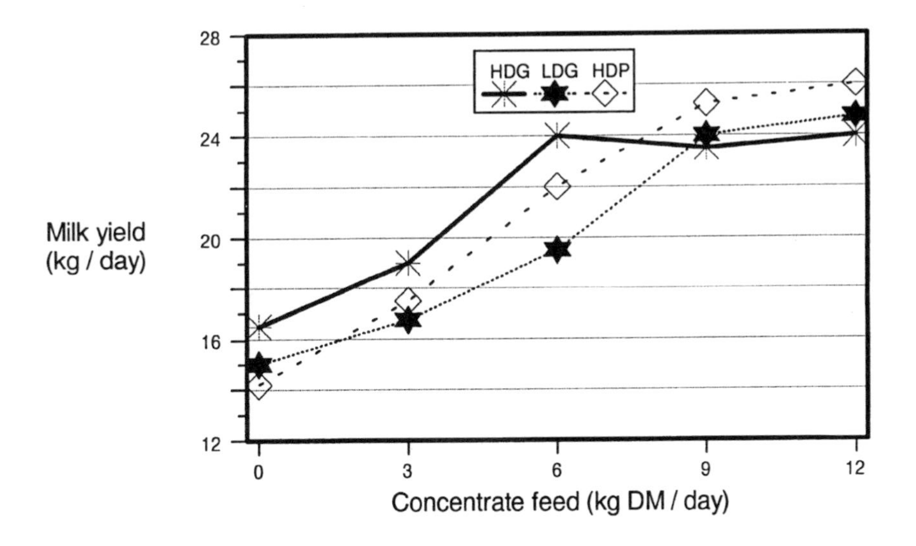

Figure 10.3 Milk production from the silages shown in Figure 10.2, at different levels of supplementary feeding (Data: MMB/IGER)

Two features of the ensilage process that appear to increase the voluntary intake of silage were described in Chapter 3, namely pre-wilting the crop to reduce its moisture content before it is ensiled, and using a chemical or biological additive to reinforce the natural ensilage process. The results of a number of experiments which have examined the effects of these two treatments on the feeding value of silage for dairy cows are summarised in Figures 10.4 and 10.5. In the case of wilting, although the dairy cows ate more dry matter of the wilted than of the unwilted silages, the unexpected result was that there was no corresponding advantage in terms of increased milk production (Figure 10.4). The reasons for this lack of response are not yet fully understood; certainly all the unwilted silages that were fed in these experiments were 'well' preserved, and the results might have been different if some of the low intake 'butyric' silages that are still frequently submitted for analysis had been included. The rate at which the crop is wilted may also have some effect. Thus in recent work at Wye College areas of the same cut crop were wilted for 48 hours, either in windrows holding three swaths, or spread over the whole area of the field, before they were ensiled. The faster wilting from the latter treatment gave a silage of 31.7 per cent dry matter, compared with 25.8 per cent from the windrowed crop; when the two silages were fed to mid-lactation dairy cows both silage intake and milk output were also higher on this faster-wilted silage (Table 10.2). Further research is clearly needed on the reasons for the lack of a consistent response by dairy cows to the higher intake of wilted than of unwilted silages.

Table 10.2 Milk production from silages made from the same crop ensiled at two levels of wilting and supplemented with 4 kg of protein concentrate per day

	Silage, slow wilt	*Silage, fast wilt*
Silage, DM%	25.8	31.7
Silage, D-value	75.9	76.5
Silage intake (kg DM/day)	13.8	15.2
Milk yield (kg/day)	24.6	26.5
Milk fat (%)	4.46	4.74
Milk protein (%)	3.18	3.23

(Data: Wye College)

Yet, despite the lack of experimental evidence, *the perception of most dairy farmers is that wilting forage crops before they are ensiled produces better feed.* Certainly the weight of crop to be handled is less with wilted than with unwilted crops; there is less risk of effluent flowing from the silo; and the silage made is more

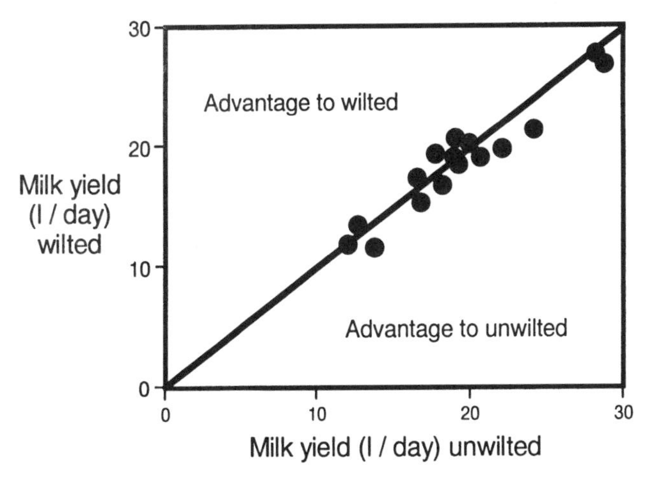

Figure 10.4 Milk production by dairy cows fed ad libitum *on silages made from the same crop with and without wilting*

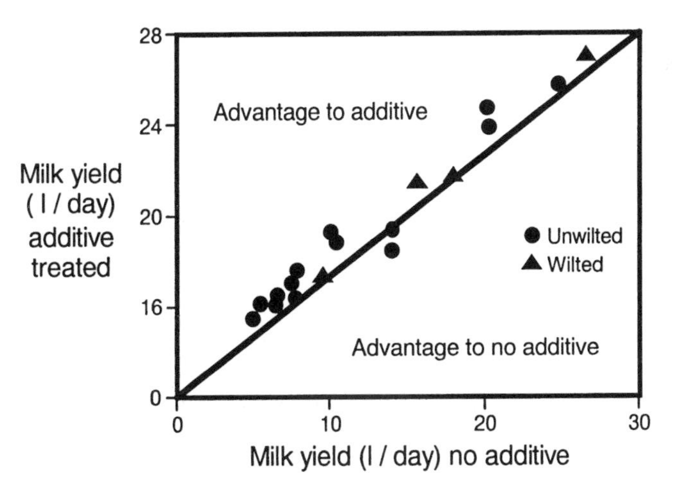

Figure 10.5 Milk production by dairy cows fed ad libitum *on silages made from the same crop with and without use of a silage additive; four of the comparisons were made with wilted silages*

'socially' acceptable. But, as noted above, the main reason may be that many *farm* silages made from unwilted, high-moisture crops are badly fermented, and so of low intake potential, whereas most *experimental* silages have been well fermented. We believe that the balance of evidence points to advantage from at least a moderate degree of wilting, carried out as fast as possible after the crop is cut.

There does appear to be a more positive response to a difference in intake between silages when this reflects an intrinsic difference in intake potential between the crops that have been ensiled. Thus Table 10.3 shows the result when silages, well made from either ryegrass or red clover, were fed as the sole forage to high-yielding dairy cows. Although the silages were of the same (rather low) D-value, the cows ate more dry matter of the red clover silage than of the ryegrass silage (reflecting the generally higher intake potential of legumes than of grasses (p. 49)), and produced nearly 2 kg more milk per day.

Table 10.3 Milk production from silages of the same digestibility, made from ryegrass or red clover

	Ryegrass silage	*Red clover silage*
Silage D-value	61	60
Voluntary intake (kg DM/day)		
– silage	8.7	10.4
– concentrate	7.0	7.0
Milk yield (kg/day)	24.6	26.5
Milk fat (%)	4.12	3.72
Milk protein (%)	2.8	2.8

(Data: GRI)

There is considerably more evidence of advantage from the increase in silage intake that can result from the use of an additive during the ensilage process, applied either because it was not practicable to wilt the crop before it was ensiled, or as a back-up, possibly at a reduced rate, when only a limited degree of wilting was practicable. The most detailed evidence has been from experiments with additives based on formic acid and formic acid/formalin, and Figure 10.5 shows the consistent advantage in increased milk production that has been found in experiments which have compared silages made from the same crop ensiled with or without one of these additives. Similar results have been reported with some other additives, including examples where the 'control' silage has been, by current standards, well preserved.

Reliable animal production data under Category A are available on about half of the additives included in the recently published Register (p. 26), and these represent a high proportion of the total quantity of additives that are currently used. However, it may be unrealistic to expect full animal production data to be produced on all of the remaining products; in that case 'approval' will be based on laboratory tests, rather than on animal experiments, with some resulting uncertainty about the practical efficacy of the materials tested. But it cannot be too strongly emphasised that, even though an additive, effectively used, may give improved silage 'quality', no additive can replace the need for a well-managed silage-making system.

The experimental work reported above has been with grass and legume silages, and a new dimension has appeared with the rapid increase in the production of wholecrop cereal silage and maize silage. Complementing the agronomic developments described in Chapter 4 there have been a number of important nutritional advances with these feeds. Thus the voluntary intake of maize silage increases markedly as the dry-matter content of the silage increases (Table 10.4) – the result either of delaying harvesting of the crop or of growing one of the new earlier-maturing maize hybrids. More significantly, and in contrast to the experience with wilted grass silages, noted above, this increase in intake when the higher dry-matter maize silages are fed leads to a considerably higher potential for animal production.

Table 10.4 Voluntary intake of maize silages of different dry-matter contents

	Silage dry-matter content			
	20	*25*	*30*	*35*
Silage dry-matter intake (kg DM/day)	10.5	11.5	13.0	15.0

(Data: IGER)

The new hybrids, and in particular those described as 'compact high-grain' types, also contain up to a third more starch than the previous later-maturing varieties, and much of this starch is retained unchanged when these hybrids are ensiled. While some of the starch in maize silage is in the form of 'digestible undegradable starch', and so does not provide as effective a source of 'fermentable metabolisable energy' (see Figure 3.4) as the starch in wheat or barley grain, maize silage does stimulate rumen protein metabolism when it is fed together with other feeds which have a high content of 'rumen-degradable' protein. Maize silage thus combines well with grass silage, in which much of the protein is in rumen-degradable form, and mixtures of maize silage and grass silage have been found to have a high potential for milk production. Experience with feeding silages made from the newer maize hybrids indicates optimum proportions in the mixed ration of about two-thirds of the forage dry matter from maize silage and one-third from grass silage (Table 10.5); higher levels of maize silage are not advised because of the risk of clinical acidosis resulting from the high starch intake.

Table 10.5 Intake and milk production from mixtures of grass silage and maize silage

Grass silage (%) *Maize silage (%)*	*100* *0*	*67* *33*	*25* *75*
Total silage intake (kg DM/day)	10.2	12.4	12.8
Milk yield (kg/day)	20.9	24.0	27.4
Milk fat (%)	4.15	4.04	3.99
Milk protein (%)	3.03	3.14	3.16

(Data: MMB/IGER)

To date the picture with 'fermented' wholecrop cereal silage, made from cereals cut at about 35 per cent dry-matter content, has been less clear, with many reports of milk output being lower than would be predicted on the basis of the estimated metabolisable energy content of the silage. Certainly the efficiency of milk production has not matched that reported from Denmark (Table 10.6), possibly because the levels of supplementary feeding (up to 8 kg of dry matter per day from barley and fodder beet) typically fed in Denmark are higher than those in the United Kingdom.

The process of conservation of wholecrop cereals by urea has been more successful under practical farm conditions, but again experience with feeding the product has been variable. One possible reason for this is that when the dry-matter content of the wholecrop cereal that is ensiled is above 55 per cent the grain may be too hard to be digested completely as it passes through the rumen – hence the advice to 'crack' the grain during the harvesting operation. At high levels of feeding the residual urea in the wholecrop silage may also place an undue burden on the protein metabolism in the rumen. Thus urea-treated wholecrop was not fed as the sole forage on any of the farms on which its use was recorded (p. 83), and in most cases it was fed in combination with grass silage, with the wholecrop fed at less than half of the total forage intake. This appeared to be successful, as the average milk yield of the recorded herds was more than 6,000 litres.

Total 'forage' intake and milk production have also been lower when grass silage has been fed as the sole forage than when it has been fed in combination with other non-concentrate feeds such as brewers grains and fodder beet (Table 10.7). One of the main advantages from the use of mixer-feeder wagons (p. 169) is that they make such mixed feeding a fully practical operation; initial research has shown benefit in both total feed intake and milk production when part of the silage in a dairy cow ration has been replaced by alkali-treated straw (Table 10.8).

Table 10.6 Milk production from wholecrop barley silage fed in mixtures with grass/clover silage and supplemented with fodder beet and protein concentrate

Silage fed	*Milk yield (kg/day)*	*Milk protein (%)*
Wholecrop alone	28.6	3.16
Wholecrop 2 Grass clover 1	26.6	3.12
Wholecrop 1 Grass clover 2	28.3	3.09
Grass clover alone	28.1	3.01

(Data: Kristensen, Denmark)

Table 10.7 Milk production from grass silage (26% DM, 14% CP) fed either as sole forage, or supplemented with brewers grains or fodder beet. Average intake data for three levels of concentrate feeding (3, 6 or 9 kg per day)

Forage fed (DM basis)	*Total forage intake (kg DM/day)*	*Milk yield (kg/day) (6 kg concentrates)*
Grass silage (100%)	10.2	20.9
Grass silage (67%) Brewers grains (33%)	12.2	27.0*
Grass silage (67%) Fodder beet (33%)	12.2	24.2

* The brewers grains contained 25% CP.
(Data: MMB/IGER)

Table 10.8 Intake and milk production over 20 weeks by dairy cows fed combinations of grass silage and sodium hydroxide-treated straw, balanced for energy and protein

Silage (%) *Treated straw(%)*	*100* *0*	*70* *30*	*30* *70*	*0* *100*
Forage intake (kg DM/day)	15.7	19.3	21.1	19.8
Milk yield (kg/day)	24.5	27.4	26.2	26.1

(Data: ADAS; Bridgets EHF)

Concentrate supplementation of conserved forages for dairy cows

The previous section has examined the effects of the digestibility and voluntary intake of silages on the level of animal production when the silages, both alone and in combination with other forages, have been fed with the same amount of feed supplement. However, in practice there are important interactions with both the amount and the type of feed supplement that is given, and these can greatly influence the production potential of different silages.

Chapter 3 showed that, as the amount of concentrate fed is increased, animals generally eat less forage; in effect the concentrate is replacing part of the forage rather than acting as a true supplement. This occurs partly because the supplement occupies some of the physical space in the rumen, but also because some supplements lower the pH in the rumen, so reducing the rate of digestion of the forage, which then passes more slowly through the rumen. As a result the production response to supplementary feeding is generally less than would be predicted on the basis of the nutrient content of the supplement.

This substitution effect is greatest when good hay is fed. It tends to be lower, but also less consistent, when silage is fed (see Table 3.6) so that the response when a supplement is fed with silage is often difficult to predict. When silage of high D-value, but whose voluntary intake is low as a result of 'poor' fermentation, is fed the amount of silage eaten may not be reduced, and may even be increased, when a feed supplement is given.

Thus in the MMB/IGER experiments already described (see Figure 10.2), the intake of the high D-value but poorly fermented silage, HDP, increased when a concentrate supplement was fed. As a result the large difference in intake between this silage and the well-fermented silage HDG, found when they were both fed as the sole feed, largely disappeared as the level of supplementary feeding was increased. As Figure 10.3 shows this had the important consequence that, at the higher levels of supplementation, there was little difference in milk production between the well-fermented (high-intake) silage and the poorly fermented (low-intake) silage; more surprisingly, the low-digestibility silage (LDG) also gave the same milk production at the higher levels of supplement feeding. The response to supplementary feeding thus varied with both the D-value and the intake potential of the silage being fed, and the differences between the silages were less marked at the higher levels of feeding; in particular, at the highest level of supplementation the intake of all the silages fell, and this led to a decrease in the efficiency of concentrate use.

The above discussion has also tended to treat the 'supplement' as a standard feed whereas, in practice, silage intake can vary greatly depending on the type of supplement that is fed. For example animals tend to eat less silage when it is supplemented with a starchy (cereal-based) supplement than when a fibrous supplement (such as dried sugar-beet pulp) is fed, probably because the fibre in the silage, and so the rate of passage of the food through the rumen, is more slowly digested under the more acid conditions in the rumen when the starchy feed is fed. In contrast, when the protein content of the supplement is increased there is a marked increase in the amount of silage eaten, and in the resulting level of milk production (Table 10.9). This effect is most marked when much of the protein in the supplement is in the form of 'rumen-undegradable' protein (p. 46), which can complement the high proportion of (rapidly) 'rumen-degradable' protein that is found in most silages. Thus fishmeal and soyabean meal, as well as a number of proprietary compounds containing 'treated' vegetable protein, have proved to be very effective supplements for silage, as also have pelleted high-temperature dried green crops (p. 47).

Table 10.9 Intake of silage and milk production by dairy cows fed silage (69 D-value) supplemented with 9 kg DM/day of concentrates containing four levels of crude protein

% CP in concentrate	*12*	*16*	*20*	*24*
Silage intake (kg DM/day)	7.44	8.12	8.65	9.04
Milk yield (kg/day)	23.9	24.9	26.0	27.1
Milk fat (%)	4.12	3.99	3.96	3.89
Milk protein (%)	3.10	3.12	3.16	3.12

(Data: MMB/IGER)

Table 10.10 Milk production by dairy cows fed silage alone, or silage supplemented with a 2:1 fishmeal: soyabean concentrate, or with a conventional 18% CP concentrate

Supplement fed	*None (silage only)*	*Fishmeal/soya (1.2 kg DM/day)*	*Concentrate (4.3 kg DM/day)*
Silage intake (kg DM/day)	11.4	12.1	10.4
Milk yield (kg/day)	15.8	20.9	21.3
Milk fat (%)	3.78	3.75	4.22
Milk protein (%)	2.95	3.17	3.07

(Data: ICI plc)

Because of the higher silage intake when these high-protein supplements are fed they are also in practice able to replace several times their weight of conventional concentrate feeds (Table 10.10). This led to the development, at the Hannah Research Institute, of a 'balancer' feed, containing 65 per cent of groundnut meal which, when fed at 0.15 kg per litre of milk as a supplement to high D-value silage has given the same milk yield as 0.4 kg of barley per litre. However, over the winter the cows fed the high-protein supplement ate some 2 tonnes more silage than when barley was fed. Whether this is economic is considered in the next chapter.

Conserved forages and milk quality

The new marketing arrangements for milk which followed the winding-up of the Milk Marketing Board in November 1994 gave added importance to the compositional quality of milk, and in particular to the possibility of manipulating its composition in response to different market requirements – and in the case of butterfat content, of avoiding exceeding the milk quota butterfat limit. Thus alongside the effect of different feeding regimes on milk production *per se*, increasing attention is now being given to the effects of feeding on milk composition, and in particular to the possibility of optimising composition for different markets.

However, an examination of the compositional data from the studies reported in the tables in this chapter does suggest that the control of milk composition is still imprecise. Milk fat content has ranged from 2.8 to 4.5 per cent and milk protein from 2.8 to 3.7 per cent, without showing any clear relation to the different feeding regimes.

In part these differences in milk composition will have been due to the different genetic potential of the dairy cows that were fed in the different experiments. However, while there are no obvious overall relationships between the make-up of the rations fed and the composition of the milk produced, a number of underlying trends can be seen.

It has long been recognised that butterfat levels in milk tend to be reduced, and protein levels increased, when a basal grass hay or silage diet is supplemented with large amounts of cereal silage or cereal-based concentrate. This is because, when these supplements are fed, they produce a fall in pH within the rumen, and under these acid conditions the microbial fermentation in the rumen produces less acetate (essential for the synthesis of butterfat) and more propionate (one of the precursors of milk protein) than at higher pH levels. Thus in the MMB/IGER study, already described (see Figure 10.2), butterfat levels below 3.5 per cent were recorded when the silages were fed alone (probably the result of a shortage of total energy in the diet) and also when the highest amounts of concentrates were fed. At the intermediate levels of concentrate feeding, however, butterfat levels were above 4.0 per cent. In contrast milk protein content rose steadily, from about 2.8 to 3.2 per cent, as the level of concentrate feeding was increased. In experiments in which energy supplements have been compared, milk fat percentage has been slightly higher with supplements high in digestible fibre than with supplements with a high content of starch, again probably as a result of the lower rumen pH when the starchy supplements have been fed.

Increasing the crude protein content of the supplement has had only a small effect on milk protein content; thus when the protein content of the supplement was increased from 12 to 24 per cent (Table 10.9) there was only a fractional increase in the milk protein percentage, although the yield of milk protein increased from 740 to 860 g per day; at the same time the milk fat content fell from 4.12 to 3.89 per cent.

The type of conserved forage fed may also affect milk composition. Thus in the feeding work reported in Figure 10.2, milk fat levels were higher when the lower D-value silage (LDG) was fed than with the more digestible silages (HDG and HDP), particularly at the intermediate levels of supplementary feeding. A similar result is seen in Table 10.1. However, the conclusion that milk fat content is higher with less digestible silages is complicated by the overall trend, in dairy production, for milk fat percentage to fall as milk yield increases; thus while the results in Table 10.1 might indicate that the lower D-value silage produced milk of higher fat content, this may have been partly confounded with the lower milk yield when that silage was fed. Similarly, in the experiment reported in Table 10.3, in which the ryegrass and red clover silages were of the same digestibility, it is not possible to attribute the higher milk fat percentage with ryegrass silage solely to differences in composition between the silages, because the milk yield was lower than from the clover silage.

Part of the fall in milk fat percentage as the proportion of maize silage fed in mixture with grass silage was increased (Table 10.5) may have been due to the higher content of starch in the maize silage; but again the result may have been in part confounded by the accompanying increase in milk yield when more maize silage was fed.

Despite this apparent confusion some broad conclusions are now accepted. When grass/legume silages of high D-value/high intake potential are fed the supplement should provide a good supply of fermentable metabolisable energy (FME), in the form of starch or sugars, both to make effective use of the highly soluble crude protein in the silage (p. 46), and also to stimulate propionate production in the rumen to balance the mainly acetate production that is likely when only silage is fed. On the other hand, when the amount of silage the livestock eat is limited, either because of restricted access or because the silage has a low intake potential, a high proportion of the FME in the supplement should come from digestible fibre, so as to maintain milk fat levels. Similarly, when the forage provides a ready supply of starch, as with silage made from the newer maize varities, or when other starchy feeds such as rolled barley, maize gluten or molasses are included in the ration, the feeding of a supplement containing digestible fibre is now recommended.

Conserved forages in dry cow feeding

In the past many dairy cows were fed at high levels of energy and protein intake during the dry period before calving ('steaming up') in order to stimulate milk production. However, this has been associated with a high incidence of milk fever, while the high body condition scores at calving (often between 3 and 4) led to considerable calving problems.

As a result there has been a marked shift in the management of the in-calf dairy cow, with most dairy farmers now carefully controlling the intake of energy during the *last month* of pregnancy so as to maintain a condition score below 3, while at the same time providing free access to palatable forage of only medium quality (60 plus D-value). Thus in the system pioneered by Somerset farmer Mike Lemmey, grazing cows are tightly stocked, at up to 11 cows per hectare, and if high D-value silage is being fed in winter the intake is restricted. Both feeds are supplemented with *ad lib* hay or straw to ensure the high total forage intake needed to develop a large rumen capacity. In addition, a special dry cow supplement is fed at 1–2 kg per day; this provides a minimum of 35 g of magnesium (which must be fed daily because it is not stored in the body), plus vitamins and trace minerals, as well as 200 g of rumen-undegradable protein, which appears to raise both milk protein production and fertility in the subsequent lactation. Practical experience is that this regime conditions the cows's rumen to deal with the high feed intake needed in early lactation, as well as giving a markedly reduced incidence of milk fever and of uterine damage and calf loss at calving.

CONSERVED FORAGES IN BEEF PRODUCTION

Systems of beef production in the United Kingdom have changed greatly over the last 50 years. In particular the huge expansion of the dairy industry in the years after the Second World War, based on Friesian cattle with their reasonable beef potential, meant that by the late 1960s some two-thirds of beef supplies were coming from the dairy herd, rather than from specialist beef herds. This preponderance of dairy-bred beef has continued, despite the more recent shift in the dairy industry to Holstein cattle, with a poorer beef potential than Friesian.

The trend away from the traditional beef breeds was also driven by the growing consumer demand for leaner meat, encouraged by medical advice to reduce the consumption of animal fat – although for many years this was contradicted by official grading standards (on which price support for beef production was based) which demanded a high level of 'finish' in beef cattle. However, market demand, greatly influenced by the requirements of the supermarkets, has increasingly been for lean cattle in the 250–280 kg carcass range, produced from calves of Friesian cows crossed with late-maturing breeds of beef cattle. The traditional early-maturing beef breeds like the Hereford and Angus have thus declined in popularity for crossing and have been replaced to a considerable extent by Continental breeds such as the Charolais and Limousin, which grow faster and give leaner carcasses (though the native breeds are now responding to the challenge!).

To exploit their higher growth potential these dairy crosses require a high level of feeding. In particular this means avoiding the winter 'store period' of most earlier feeding systems, during which cattle were expected to make little liveweight gain during the winter – and for which hay, silage and straw of only 'maintenance' quality were quite adequate feeds.

Because of the generally poor quality of the hay and silage that was available on most beef farms, the initial response to this demand for better winter nutrition was to feed increasing amounts of cereals to beef cattle – leading to the development, at the Rowett Research Institute in the early 1960s, of the 'barley-beef' system, noted in the Introduction, in which beef cattle were fed just on barley and soyabean meal. Cattle fed in this way made rapid rates of gain; yet barley-beef feeding did not become the dominant system of beef production in the United Kingdom that was predicted at the time. For although barley-beef has continued to supply a specialist premium market, its main effect was to stimulate the development of alternative feeding systems, which would give faster and more consistent gains than in traditional beef systems. These systems aimed to feed less cereals – then becoming a high-cost feed resource within the EU – as well as producing heavier carcasses than when only barley was fed, so spreading the initial cost of the beef calf over more units of output.

A key feature of these systems, in which beef cattle were slaughtered at between 14 and 24 months of age, in contrast to the 24 to 36 months of earlier production systems, was that the animals had to make good daily gains throughout their life, both summer and winter, with only limited use of cereal feeds. In summer this demanded improved grazing management, including topping to maintain the digestibility of the grazed pastures, and in winter the feeding of more productive conserved forages.

Yet, while there were significant improvements in the quality of the hay and silage that were fed to beef cattle, improved methods of forage conservation were not adopted as widely, or as quickly, on beef farms as on dairy farms – at least in part because the generally smaller size of beef than of dairy enterprises meant that it was more difficult to justify the levels of investment in equipment and management skills that were being adopted in the dairy industry. However, this situation changed with the introduction of big-bale silage in the 1980s, both because this method of conservation allowed both large and small lots of grass to be ensiled with low losses, and because big bales proved a very effective way of feeding silage to groups of beef cattle, either in yards or in the field.

Much of the big-bale silage fed on beef farms is now made by contractors, in a more efficient and economical operation than would be possible with much of the available 'on-farm' harvesting equipment. The silage, distributed using an automatic unloader (see Plate 9.2), has proved particularly useful in the feeding of suckler beef cows, most of which are kept outdoors throughout the year, and of yarded cattle, and increasing amounts of silage are now being fed in both specialist and secondary beef enterprises – though there is some evidence that high-silage diets may be increasing the fat content of beef carcasses, against the market trend. Certainly in the higher-rainfall, mainly grass regions of the

United Kingdom, in which haymaking is a less reliable process, ensilage is becoming the preferred method of forage conservation – though many beef farmers will still aim to make some hay for their young stock.

The importance of forage quality

For a beef animal to make good liveweight gains it must have a high intake of nutrients – that is, it must be able to eat large amounts of food of high digestibility. The importance of high digestibility was shown in the work reported in Table 3.5, in which daily gains were lower on the two later-cut chopped dried grass feeds because both D-value and voluntary intake were lower than on the early-cut feed. Similar benefits from earlier cutting can be expected with hay; thus in one experiment cattle fed on ryegrass hay of 70 D-value gained 0.9 kg per day, compared with only 0.7 kg by similar cattle fed hay made from the same crop, but cut four weeks later, and of 60 D-value.

Research has also shown the benefits from earlier harvesting for silage (e.g. Table 3.3), but with the vital qualification that the silage must also have a high intake potential. For voluntary intake, and the resulting daily gains, can be very low if the silage is badly fermented and, as with dairy cow feeding, silage intake is likely to be at least as important as silage digestibility. Thus in work at the Northern Ireland Agricultural Research Institute well-fermented silage, made with formic acid (pH 4.2 and low ammonia content), had an intake of 8.5 kg dry matter per day and gave daily gains of 0.89 kg, compared with an intake of 6.3 kg and daily gains of only 0.47 kg, when the silage fed was made from the same crop, but badly fermented (pH 4.8 and high ammonia content).

This has been confirmed in other experiments, in which silages made with an additive have given higher intakes, and an average increase in daily gain of 0.18 kg when compared with similar silages made without an additive (Figure 10.6). In most of these studies the additive used has been based on formic acid, but recent work has shown similar advantage in both intake and daily gains by beef cattle when a biological additive has been used (e.g. Table 10.11).

As was shown in Chapter 2, the voluntary intake of silage is also increased if the crop is wilted before it is ensiled, and a limited number of experiments with beef cattle have shown daily gains of the order of 0.12 kg per day higher when wilted silages have been compared with unwilted silages. Thus wilting, which also reduces

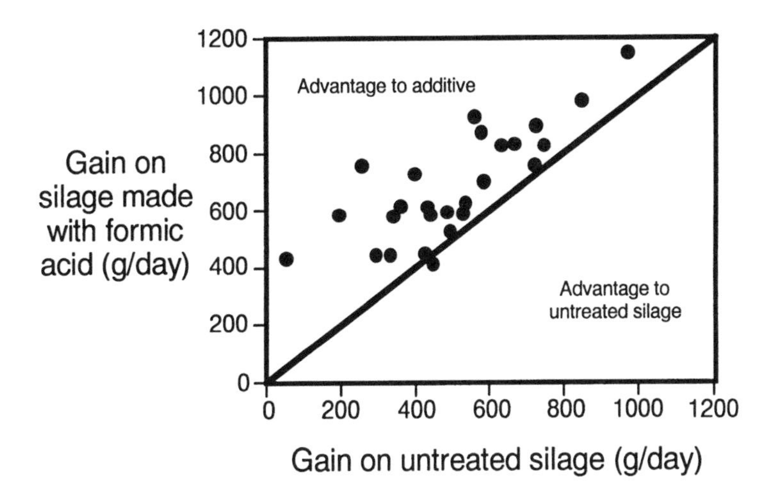

Figure 10.6 Daily liveweight gains by beef cattle fed ad libitum *on silages made from the same crop with and without use of a silage additive*

Table 10.11 Daily liveweight gains by cattle fed silage made from grass of 15% DM content with no additive, with a formic acid additive, and with a biological additive

Silage	*No additive*	*Formic acid additive*	*Biological additive*
Silage intake (kg DM/day)	6.3	7.0	6.7
Diet D-value	64	67	67
Liveweight gain (kg/day)	0.82	0.97	1.00

(Data: Hillsborough Agricultural Research Institute, N. Ireland)

the risk of effluent flow from the silo, should be carried out whenever conditions permit.

The feeding of cereal and energy supplements

In a number of the experiments reported in the previous section beef cattle, fed only on hay or silage of high D-value, made gains of the order of the 1 kg per day needed in modern beef production systems. However, it will seldom be possible, under practical farming conditions, consistently to make large amounts of hay or silage of this quality. Thus many lots will be of lower value, and must be supplemented with other feeds in order to achieve the required rates of gain, particularly during the later stages of feeding.

The most common deficiency when hay or silage is fed is in energy intake (low intake and low D-value), and this is generally made good by feeding an energy supplement such as rolled barley, maize gluten or distillers by-products. However, for the reasons noted on p. 53, when energy supplements are fed the stock may eat less forage; as a result the response to supplementary feeding is often less than would be predicted, particularly with hay and silage of high D-value. Thus in one experiment a supplement of 1 kg of cereal gave close to the expected extra 0.25 kg daily gain when it was fed with silage of 58 D-value, but only an extra 0.1 kg per day with a higher-quality silage of 65 D-value. Energy supplements may also reduce the intake of silage made from a wilted crop or with an additive more than with the corresponding untreated silage.

This means that part of the advantage from earlier cutting for silage, and from wilting or the use of an additive, may be lost when an energy supplement is fed. In that case, in terms of the compromise between 'yield' and 'quality', presented in Figure 3.1, the date at which forage is cut and conserved for beef production should probably aim for a high yield of intermediate D-value material (about 65 per cent), rather than for a lower yield of silage of high D-value. However, as discussed in the following section, when a protein supplement is fed the reduction in forage intake may be less than with a straight energy supplement, so that the benefits from earlier cutting can be more fully exploited.

Protein supplements for beef cattle

As noted in Chapter 3, high protein feed supplements tend to reduce the intake of silage less than when energy supplements are fed, so that the forage can then contribute more to the total ration. However, while cattle of all ages will benefit from the higher forage intake, younger cattle will *also* get nutritional benefit from the supplementary protein. Thus, while the protein requirements of older cattle can often be supplied from the forage alone, hay or silage fed to young cattle is often deficient in protein, particularly so with silage in which much of the protein is highly soluble and so is inefficiently utilised in the rumen. As a result the most marked responses to the feeding of protein supplements have been with young beef cattle. Recent work at IGER has compared fishmeal and rapeseed meal as supplements, at 10 per cent of the weight of silage dry matter, and fed with two silages of 60 and 70 D-value to beef cattle from 120 to 230 kg liveweight. When the silages were fed alone, average daily intakes were 21.8 and 25.5 g of dry matter per kg liveweight, and daily gains were 0.56 and 0.69 kg per day with the low- and high-digestibility silages. When the supplements were fed average intake on the low-digestibility silage increased to 23.5 g dry matter per kg liveweight, and daily gains to 0.68 kg, and on the higher-digestibility silage to between 23.8 and 25.4 g dry-matter intake with daily gains of 1.0 kg. No difference was found in response between the two protein supplements, almost certainly because they both contained a high proportion of the 'rumen-undegradable' protein that was needed to complement the highly degradable protein in the silages.

Table 10.12 shows the results of two feeding experiments in Northern Ireland, in which cattle of about 130 kg liveweight were fed on silage of 70 D-value supplemented either with feed containing the same amount of crude protein from soyabean, fishmeal or maize gluten, or with two levels of barley. Silage dry-matter intake and daily gains were consistently higher with the protein

Table 10.12 Daily liveweight gains by cattle fed 70 D-value silage plus supplements of barley, barley + soyabean and barley + fishmeal

Supplement (kg DM/day)	*Barley 1.4*	*Barley 2.25*	*Barley + soyabean 1.4*	*Barley + fishmeal 1.4*
Trial 1				
Silage intake (kg DM/day)	2.31	1.61	2.50	2.53
LW gain (kg/day)	0.84	0.95	0.98	1.01
Trial 2				
Silage intake (kg DM/day)	2.38	1.77	2.53	2.54
LW gain (kg/day)	0.90	0.99	1.04	1.03

(Data: Hillsborough Agricultural Research Institute, N. Ireland)

supplements than with the barley. The advantages of earlier cutting to give more digestible forage are also likely to be better exploited when protein supplements are fed.

CONSERVED FORAGES FOR SHEEP

Because most sheep have grazed outdoors during most of the year, conserved forages have traditionally played a smaller role in the feeding of sheep than in either dairy or beef cattle feeding. Supplementary feeding, generally with low-quality hay or straw plus limited concentrates, has been confined to ewes during late pregnancy and lactation. Other sheep have been fed only when there has been an acute shortage of grazing, as when the ground has been snow-covered.

The prevailing move towards more intensive management of sheep has led to higher summer stocking rates on both upland and lowland pastures, and this has resulted in shortages of forage for grazing during the winter and early spring. This shortage has been accentuated by higher lambing percentages – and so higher nutritional demands on the ewe – during the same period, and overall there has been a greatly increased requirement for supplementary feeding during the winter. This has become more practicable as a result of the wide adoption of winter housing for sheep, which has simplified the operation and control of hand-feeding; but it does mean that provision has to be made for more of the winter feed requirements of the flock.

While the costs of hay, silage and straw, relative to cereals, may not be as attractive as in the past, they do still provide the cheapest source of metabolisable energy on most grassland farms and so should supply as much as possible of this supplementary feed. However, the provision of enough conserved forage for winter feeding does pose some problems. Thus on lowland farms the pressure is to move the ewes and lambs from indoor winter feeding out on to pasture as soon as enough grazing is available, so that their increasing daily feed requirements match the speeding-up of pasture growth during April and May (see Figure 4.1). At the same time the aim is to cut part of the grassland area to provide hay or silage for the following winter, with the regrowth providing the 'clean' grazing, free from parasitic worms, needed by the lambs after weaning.

However, the currrent trend towards earlier lambing is putting increasing demands on the provision of grazing in the spring and, although more nitrogen is now being applied to sheep pastures than in the past, enough forage may not be available for both grazing and cutting at this time of year. Thus increasing importance is being given to cutting for conservation later in the year, when many of the lambs will have been sold, and when the grazing demands of the ewes are low so that they can be stocked more heavily. Much of this mid- and late-season grass becomes available in relatively small lots, which can be made into hay more readily than heavy spring crops; these smaller lots are also well adapted to storing as big-bale silage, often made by a specialist contractor. Regrowths from these later cuts can then provide the highly digestible grazing needed to flush the ewes in autumn.

Sheep on the hills and uplands pose greater problems. Previously it was most often the winter carrying-capacity that determined the number of sheep that could be carried on upland farms but, with greater use of winter housing, it is more often the amount of forage that can be stored for winter feeding that is now the limiting factor. Because it may not be possible to conserve enough feed for the numbers of sheep that can be grazed during the summer, supplementary feed must be brought in. ADAS trials have shown that in many cases it may then be more economical to purchase straw rather than hay, for the daily cost of feeding a ewe in winter on straw plus 1.0 kg of concentrates may be less than the cost of the hay plus the 0.7 kg of concentrates needed to provide the same amount of energy.

All bought-in feeds are expensive, and on most sheep farms the aim must be to produce as much as possible of the conserved winter forage that will be needed. Traditionally this forage has been conserved as hay, reflecting the widely held view that silage is not a suitable feed for sheep. It is now clear that this was more a reflection of the poor quality of most of the silage that was made on sheep farms, rather than of any basic deficiency in silage as a feed for sheep. With the wider adoption of improved silage techniques, and in particular of big-bale silage, excellent results are now being obtained with feeding silage to sheep.

This is fortunate for, although recent changes in support regimes have led to greatly increased numbers of sheep being kept in lowland areas, most of the sheep in the United Kingdom are still in the upland areas in the north and west of the country, where silage has advantage over hay as the main method of forage conservation. The practical success that has been achieved from feeding well-fermented silage to sheep is thus of great importance. There will, of course, be advantages if the cut crop can be partly wilted before it is ensiled. But wilting just cannot be relied on in those areas of the country where most sheep are congregated, and there is then a very strong case for using a silage additive, if necessary at an above-average rate, to ensure the good fermentation needed to get a high intake when the silage is fed to ewes.

As has been noted, much of the silage made on both upland and lowland farms is likely to be in relatively small lots, and there has been a remarkable expansion in the production of big-bale silage for feeding sheep, both for these smaller lots and for the main silage harvest. Silage stored in bales is also very convenient for feeding to groups of sheep, both indoors and outdoors. One disadvantage of baled silage made with the original equipment was that ewes often had difficulty in eating enough silage because the grass was not chopped before it was baled; this difficulty has been overcome with the newer equipment for baling chopped grass, described on p. 151. The benefit from feeding silage made from chopped grass has been shown in recent work in Northern Ireland, in which ewes fed on silage made from chopped grass gave higher lamb birthweights than ewes fed on flail-harvested grass (Table 10.13). It is also important to ensure that the ewes have ready access to the silage, yet without wastage.

Table 10.13 Intake of silage by ewes during pregnancy, and resulting lamb birthweights, in a comparison between silages made from flail-harvested and precision-chopped grass

Type of silage	*Flail*		*Chop*	
Silage intake in mid-pregnancy (kg DM/day)	1.05		1.39	
Concentrate given during last six weeks of pregnancy (kg)	6	18	6	18
Silage intake in late pregnancy (kg DM/day)	0.66	0.84	1.09	1.06
Lamb birthweight (kg)	4.80	5.20	5.10	5.40

(Data: Hillsborough Agricultural Research Institute, Northern Ireland)

Ewes are likely to remain at pasture for the first two or three months of pregnancy and, when first housed, they require feed of only medium quality so that they do not become overfat. Hay or silage of quite low D-value – in the 50–55 range – possibly supplemented with a feed block containing non-protein nitrogen such as urea should be fully adequate. Alkali-treated straw (p. 167) can also be fed at this time, although some ewes do not eat this feed very readily. During the last two months of pregnancy forage quality is more critical, because the ewe's capacity for feed becomes increasingly restricted by the volume of the foetus, just at the time that her nutrient requirements are increasing. Thus to keep the amount of concentrates that are fed at a reasonable level the hay or silage must be of higher D-value.

Under experimental conditions the full nutrient requirement of a ewe carrying and suckling twin lambs can be provided by hay of 77 D-value. But to make enough hay of this quality, even with barn-drying facilities, would in general be quite impractical. High D-value silage is also unlikely to be an adequate feed, and a more realistic target on most farms would be for hay or silage of around 65 D-value; supplemented with 450 g of concentrates per day, forage of this quality should provide fully adequate feed for a ewe carrying twin lambs – provided, if silage is being fed, that it has been well fermented and that the method of feeding does not restrict intake.

If the conserved forage is of lower digestibility correspondingly more supplementary feed will be needed, and this is likely to reduce the amount of forage the ewe is able to eat; thus when hay of only low D-value is available as much as 740 g per day of high-energy concentrate may be needed (Table 10.14). This high

Table 10.14 The intake of hay and silage of different digestibilities, and of concentrates, eaten by ewes during the last five weeks of pregnancy. The calculated ME contents of the diets, in relation to the daily requirement of 16 MJ of ME, are also shown

Conserved forage	*Hay*	*Hay*	*Hay*	*Silage*
D-value of forage	77	64	50	68
Intake (g DM/day)				
– forage	1630	1115	560	1190
– concentrate	0	370	740	370
Estimated ME in diet (MJ/day)	19.8	16.0	13.6*	17.0

* This diet would be deficient in energy.
(Data: AGRI)

level of supplementary feeding is costly, and may also lead to digestive upset; thus every effort must be made to conserve more digestible forage. There is also increasing interest in the feeding of specially formulated feed blocks during late pregnancy and lactation. While these may be more expensive than conventional concentrates, most feed blocks contain rumen-undegradable protein, which is claimed to increase the amount of conserved forage that the ewes will eat, and so reduce overall feed costs.

Once the lambs are born the ewe's intake of forage will be less limiting than during pregnancy, because of the effective increase in the capacity of her rumen. Thus if several lots of hay or silage of different digestibilities are available for winter feeding the best lots should be allocated for feeding in late pregnancy, when feed intake is most limiting. But to keep overall concentrate feeding below about 500 g per day a forage quality of above 60 D-value should be aimed for during both late pregnancy and lactation, and wherever possible the forage should be short-chopped.

This chapter has examined some of the technical possibilities for milk and meat production from rations containing a high proportion of conserved forage, and has shown the high levels of animal production that can be obtained. Less attention has been given to the factors in different feeding systems that may determine their relative profitability under a range of different farming situations, for a system that places maximum reliance on the feeding of conserved forage will not necessarily be the most profitable. For example, the high winter liveweight gains that were recorded when a protein supplement was fed to beef cattle (Table 10.15) would probably have been profitable if the cattle were to be sold directly for slaughter: they might not have been worthwhile if the cattle had to be turned out for a further period of grazing before they were marketed, because unsupplemented cattle, which have made lower winter gains, can quickly make this up through cheap 'compensatory growth' at pasture.

Thus the following chapter examines some of the management and economic factors which must be considered in deciding the optimum forage conservation strategy for different livestock enterprises – as well as the possible impacts of the rapidly changing but uncertain political, economic and environmental restraints under which livestock farmers may have to operate in the future.

Table 10.15 Daily liveweight gains during winter (184 days) by yearling cattle fed silage *ad lib* plus 1 kg/day of either barley or a protein supplement

Supplement	*Initial LW (kg)*	*Final LW (kg)*	*Daily LW gain (kg)*
Barley	210	288	0.4
Barley 40% Soyabean 60%	218	394	0.9
Barley 75% Fishmeal 25%	220	400	0.95

(Data: ADAS Northern Region)

CHAPTER 11

FORAGE CONSERVATION IN FARMING SYSTEMS

In the last edition of this book, published in 1986, we wrote of the 'considerations that make plotting the future course of forage production in United Kingdom agriculture so difficult'. We also noted that 'it is because of the *uncertainty* about the future framework within which agriculture will have to work, even a few months ahead, that *Forage Conservation and Feeding* has dealt mainly with technical options, rather than seeking to give specific advice'. Little has changed. The future is at least as uncertain today as it was then – as is illustrated by two leading articles in the farming press in spring 1995. One examined the relative 'effectiveness' of five different ways of *reducing* milk production, as an alternative to buying quota at 50p per litre so as to avoid paying the superlevy on above-quota milk output. (Similar articles appeared in spring 1996, except that milk quota was by then nearer 70p per litre.) The second, from ADAS, considered the new situation in which wholecrop wheat, barley and oats cut for silage would continue to get support under the Arable Area Payment Scheme, whereas support for forage maize would be drastically reduced, as a result of the overshoot in the area planted to maize in 1994. Neither article had anything to do with improving the technical efficiency of milk production.

The UK livestock industry faces other uncertainties. Thus milk prices, following the demise of the Milk Marketing Board in November 1994, have risen sharply, yet most dairy farmers question whether these higher prices will be sustained in the longer term. Meat production is supported by the Sheep Annual Premium, Beef Special Premium and Suckler Cow Premium Schemes (as well as by Hill Livestock support); but these are increasingly aimed at discouraging more intensive management, and in some areas restrictions on stocking rates are already being introduced so as to prevent 'environmental overgrazing'.

On the other hand, cereal prices are higher than in 1992, despite the assumption in the 1992 review of the Common Agricultural

Policy that by now they would have fallen below £90 a tonne – a profitable situation for cereal growers (Table 11.1), but less so for livestock farmers, for whom cereals represent a major input cost. At the same time the live animal export trade, which grew rapidly during the 1980s, and which has made an important contribution to the incomes of both sheep and dairy farmers, is being vigorously challenged by animal welfarists and by the impact of the BSE crisis, which has arisen since this chapter was written. Perhaps more seriously, the concern for animal welfare is being exploited by the vegetarian movement, as part of its longer-term assault on the whole concept of livestock farming.

Table 11.1 The returns for feed wheat, predicted following the 1992 reform of the Common Agricultural Policy, compared with the actual returns in 1994 and 1995. (Average yield 7.25 tonnes per hectare)

	1994 (predicted)	*1994 (actual)*	*1995 (actual)*
Average market price (£/tonne)	83	103	115.50
Area payment (£/tonne)	22	27	37.00
Total return (£/tonne)	105	130	152.50
Extra return (£/ha) over predicted return	–	180	345.00

(Data: Nix, Wye College, University of London)

All this will be history to future readers of this chapter. They will know what happened next – and will themselves be facing new problems! Yet, despite these uncertainties, there do appear to be a number of underlying trends which could be important in the context of forage conservation, the subject of this book. When we wrote the first edition, in 1972, we noted that, although many enthusiasts were getting high animal output from their grassland, they would probably 'have had less bother – and made more money – by increasing their summer stocking rates, conserving less grass, and buying in more subsidised concentrates'. This situation changed after the United Kingdom became a member of the EU, and even more so with the introduction of milk quotas, for dairy farmers now had a new target, 'to produce their quota as cheaply as possible'. With the milk to concentrate price ratio in 1984 down to 1:1, grass, both grazed and conserved, was certainly the 'cheapest feed' and, as a result, much greater quantities of grass and forage crops were conserved for winter feeding (see Figure 1.2). This set in train a remarkable change in the management of grassland in the United Kingdom. Thus, according to Roger Wilkins of IGER, 70 per cent of the grass that was grown in 1980 had been grazed while only 30 per cent was cut for conservation; in contrast, by 1993, not only was more grass being grown, but the proportion cut for conservation had risen to 50 per cent.

There are several reasons, however, why this situation must be reassessed in 1996. While returns from milk have risen considerably the 'real' price of feed grains (despite the figures shown in Table 11.1) is now lower than in 1984, when milk quotas were imposed. As a result the milk to concentrate price ratio has risen to 2.25:1, and seems likely to remain above 2.0:1 for the foreseeable future (for if milk prices fall, so also will the price of cereals, if CAP reform finally begins to be effective).

Thus feeding cereals and other energy concentrates to dairy and beef cows is now more profitable than it was ten years ago. In contrast, over the same period, conserved grass has become considerably more expensive, as a result of higher seed and fertiliser prices, more expensive machinery and labour and, in the case of silage, the considerable costs of new measures to prevent effluent pollution. As we discuss below, grass, efficiently grazed, is still the cheapest feed for ruminant animals; but while costs do differ from farm to farm, the cost of energy from conserved grass is now between two and three times the cost of energy from grazed grass – and in some cases is as high as the cost of energy from cereals and feeds such as maize gluten. More controversially, on many farms silages made from forage maize and wholecrop cereals may provide a cheaper source of feed energy than grass silage; thus in 1994 ADAS calculated that a GJoule of Metabolisable Energy from grazed grass, wholecrop cereal silage, maize silage and grass silage cost £1.9, £4.6, £5.1 and £5.8, respectively (excluding Area Support which, in 1994, further reduced the costs of wholecrop and maize silage to £3.5 and £4.5 per GJoule of ME).

As we have noted, the initial response of most dairy farmers to the imposition of milk quotas was to keep the same number of cows, but to reduce the milk yield from each cow by feeding fewer concentrates and more grass and conserved forages; this was encouraged by the good returns from the sale of surplus calves.

This strategy is changing in the new situation of higher hay and silage costs, the more favourable milk to feed price ratio, and the possibility of lower calf prices if exports are restricted. Thus, as Figure 1.1 showed, average milk yield per cow is once again rising, as is average herd size, reflecting the marked reduction in the number of dairy enterprises (predicted to fall by a further 30 per cent over the next decade). This is being accompanied by a number of other important developments.

The increasing costs of forage conservation (including the recent 40 per cent rise in the cost of bale-wrapping film) is stimulating a renewed interest in the potential for making greater use of grazed grass, in particular early and late in the year. The idea of 'extending the grazing season' is, of course, not new; but research, notably at the Hillsborough Research Institute in Northern Ireland and at the Scottish Agricultural College at Crichton Royal, is examining the novel concept of giving dairy cows limited access to grazing *while they are still being fed mainly on silage* – in spring before enough grass has grown to permit full grazing, and in late autumn, when there is often plenty of grazing available but it is only of moderate quality. To date the main benefits appear to have been from extended autumn grazing, and both centres have found that milk yield has increased when dairy cows have been given restricted access to autumn pasture in addition to their main feed of *ad lib* silage; the cows have also eaten less silage, so that less silage needs to be stored for the full winter feeding programme. Table 11.2 shows results from Northern Ireland, with both spring- and autumn-calving cows giving higher milk yields when they

Table 11.2 Benefits from giving silage-fed dairy cows access to grazing for 2–3 hours daily from 29 October to 26 November

	Spring Calvers (Silage + 2 kg cake)		*Autumn Calvers (Silage + 6 kg cake)*	
	Housed	*Housed + grazing*	*Housed*	*Housed + grazing*
Silage intake (kg DM per day)	10.7	6.7	11.0	6.8
Milk yield (l per day)	12.3	14.7	23.1	25.2
Milk fat (%)	4.18	4.27	4.12	4.00
Milk protein (%)	3.22	3.46	3.12	3.27

(Data: Hillsborough Agricultural Research Institute, Northern Ireland)

were strip grazed for two to three hours, after morning milking, on paddocks that had regrown for 60 days after being cut for second-cut silage. Similarly, at Crichton Royal autumn-calving cows have yielded up to 25 litres of milk daily on silage plus only 1 kg of concentrates, when they have also had access to grazing up until early December.

Clearly early and late grazing will involve some additional labour in moving the herd to and from the field each day, and in moving electric fencing – but it will also reduce the amount of slurry that has to be dealt with at the end of the winter. It could well be more widely adopted.

'Extending the grazing season' is only one aspect of a wider re-examination of the integration of cutting and grazing in the management of grassland, aimed at dealing with the uneven growth of grass at different times of year (see Figure 4.1). Thus once it is accepted that grazing alone need not provide the full daily intake of forage, stock can be turned out in the spring on to part of the grassland area that has had an early application of N fertiliser (with timing based on the T-sum, p. 71) up to two weeks earlier than if the grazing is required to provide the full feed – surely the correct use of the term 'early bite' (though it may be difficult to combine this with late autumn grazing, which will almost certainly delay grass growth in the spring). Then, once sufficient grass is available for full-time grazing, it is now accepted that both dairy and beef cattle can be stocked more heavily than has generally been practised in the past, yet without loss of individual animal output, because of the high digestibility, palatability and freedom from soiling of the grass grown during the first part of the grazing season. Decisions on the optimum stocking rate and the duration of grazing can then be based on techniques of sward measurement such as those recommended by ADAS since the late 1980s. This permits heavier stocking on the grazed area shown in Figure 4.2, so releasing a bigger area of grass to be cut for first-cut silage – an advantage because of the uniquely superior quality of first-cut silage compared with later cuts. Further, the autumn 'surplus' of grass (see Figure 4.1), which might previously have been cut for silage at the end of October, can instead be grazed as a supplement to indoor silage feeding.

Figure 4.1 also shows that the rate of growth of grass falls in mid-summer. This happens not just in the lower-rainfall areas of the United Kingdom, for there can be a shortage of forage for grazing after mid-summer in most areas, particularly in a dry year and where stocking rates are high. This has led to renewed interest in the role of 'buffer feeding' – *giving additional feed to grazing animals whenever they cannot get enough feed from grazing alone.* In the past this has generally been done by feeding extra concentrates, although left-over silage (often of pretty poor quality) from the previous winter has sometimes been fed.

However, recent research has aimed to make buffer feeding a more positive component of grassland management, and less of a salvage operation. This has become more practical with the wider availability of big-bale silage, which allows silage of good quality to be fed to grazing livestock without the need for opening up a main silage clamp in mid-summer, with the risk of high wastage at the silage face. This silage *can* be fed from racks in the field, but it is easier to monitor and to control silage intake if it is fed from troughs 'indoors' at one or both milking times (the alternative, of self-feeding from a clamp silo, risks high wastage due to moulding). Research in Scotland has shown that grazing dairy cows, given either restricted or unlimited access to silage, required considerably less grazed grass. There was little effect on the milk production of these late-lactation cows (Table 11.3), but the silage-supplemented cows gained more liveweight.

Table 11.3 Benefits from buffer feeding late-lactation dairy cows with silage, either with 45 minutes' access per day, or *ad libitum*, during August and September

	Grazing only	*Grazing + 45 minutes silage*	*Grazing + ad libitum silage*
Grazing intake (kg DM/day)	10.2	7.4	2.6
Silage intake (kg DM/day)	0.0	4.1	10.4
Concentrates (kg DM/day)	2.4	2.3	2.6
Milk yield (kg/day)	13.7	14.3	14.1
Milk fat (%)	4.13	4.14	4.41
Milk protein (%)	3.87	3.87	3.79

(Data: Milk Marketing Board/West of Scotland College, Scottish Agricultural Colleges)

There is also considerable interest in the potential of other feeds, including alkali-treated straw (p. 167), wholecrop silage (possibly cut in late July for buffer feeding a few weeks later in early autumn), brewers grains and citrus pulp, all of which may be available more cheaply than grass silage.

This search for cheaper feeds reflects the continuing trend for the costs of 'grass' silage to rise, particularly in comparison with silage made from forage maize and wholecrop cereals, noted above. Thus, as well as 'extending the grazing season' in order to reduce the amount of forage that has to be conserved for winter feeding, there could also be a gradual shift, both for winter feeding and for buffer feeding, from 'grass' silage to 'cereal' silage. Cutting 'surplus' grass for conservation in the spring will continue to be an essential part of good grassland management, but there may be less emphasis on intensive management of grassland to provide forage for later silage cuts – although storing grass not needed for current grazing, generally as big-bale silage, will continue to be a key part of overall grassland management.

Nevertheless, large amounts of grass will still be conserved as silage, and there is thus particular interest in the remarkable animal responses that have been found when grass silage and maize silage are fed in combination, as a way of exploiting the particular nutritional features of both feeds. In our view similar research on the combined feeding of grass silage and wholecrop cereal silage, both fermented and urea-treated, is urgently needed.

This opens up the whole subject of 'complete diet' feeding, and the operational role of complete diet feeding systems (p. 183). For, although mixer-feeder wagons have been extensively used on farms since the mid-1970s, there is still controversy as to the economic case for the additional capital and labour costs involved in complete diet feeding, compared with easy-feed or self-feeding systems. In particular there has been little research on the possible differences in animal production between rations fed in mixture or as separate components, though recent work at the Hillsborough Research Institute has shown that dairy cows fed on a complete diet gave 3 kg more milk per day than cows fed the same ration components separately. However, as animal feeding becomes more precisely controlled, with the aim both of improving the nutritional balance of the ration and of reducing costs, it may be increasingly difficult to achieve this under a system of self-feeding, in which many of the decisions on how much of the feeds on offer are eaten are taken by the animals that are being fed, rather than by the livestock feeder. Thus as dairy herd size increases, and particularly with higher-yielding animals, there is likely to be a gradual shift from self-feeding to controlled feeding systems. The elite dairy herd at ADAS Bridgets Dairy Research Centre may still be atypical, but it is difficult to imagine the

10,000+ litre herd of the future being fed the Bridgets ration of 26.25 kg of maize silage, 5.25 kg of grass silage, 1.5 kg of molassed sugar-beet feed, 3.5 kg of rolled wheat, 4.0 kg of wheat Sodagrain, 0.75 kg of ground maize, 0.75 kg of fishmeal, 1.0 kg of heat-treated soya, 3.5 kg of soyabean meal, 0.15 kg of Megalac and 0.2 kg of minerals *by any other than a complete diet feeding system*!

An extension of this concept is to the system of 'storage feeding', in which dairy and beef cattle are fed indoors all the year round on stored conserved forages, cereals and compound feeds. In the United States many milk production units with more than a thousand cows now operate with storage feeding, coupled with increasingly automated milking parlours. However, we think it unlikely that this system will develop on any scale in the United Kingdom, because it involves cutting and storing the whole forage requirement, with the high costs of this operation, while at the same time not exploiting the considerable contribution that grazing still offers to the ruminant livestock economy. But the precise balance between grazing and cutting of grassland, between grass hay, grass silage, and maize and wholecrop silage, is still uncertain – and is likely to remain so until the paradox of future cereal prices, spelt out in Table 11.1, is decided.

This is just one aspect, albeit a key aspect, of the uncertainty over the future direction of the Common Agricultural Policy, which continues to dominate agriculture in the United Kingdom and Europe, and to have repercussions throughout the rest of world agriculture. According to Brussels no change in that policy is needed; but the report of the UK Minister's CAP Review Panel, published just before we went to press, made it clear that the 1992 reform of the CAP, already inadequate, will be quite unable to cope with the planned eastward enlargement of the Community. What is less clear is how that 'reform' should be further reformed. Should the 'market' rule, as advised by the economists on the Panel, with the inevitable polarisation (and not just in the UK), into regions of high-tech, intensive farming, and other regions farmed either at low intensity or abandoned? Would the latter include much of the 'grassland' area of the country, as feared by some other members of the Panel? What would the 'non-market' consequences of such a policy be on rural society and the rural environment? And could there be some return to 'mixed' farming, with the reintroduction of grassland on farms which at present grow only arable crops?

As we speculated at the end of Chapter 4, the new importance being placed by contemporary society on the rural environment could also mean that 'environmental conservation', rather than 'forage conservation', will become the priority in the management of much of the grassland area in this country – and that as a result much of the current funding for area and livestock support will be transferred to environmental support. The key question, then, would be how best to do this while still retaining a vigorous and viable livestock industry in the UK. Hopefully we may have some of the answers to these questions in time for the next edition of *Forage Conservation and Feeding*!

ABBREVIATIONS USED IN TEXT AND TABLES

ADAS	The Agricultural Development and Advisory Service (England and Wales)
AFRC	The Agricultural and Food Research Council
AGRI	The Animal and Grassland Research Institute (formed by the amalgamation of GRI and NIRD)
CAP	The Common Agricultural Policy of the European Union (EU)
CEDAR	The Centre for Dairy Research, University of Reading
DANI	The Department of Agriculture for Northern Ireland
EHF	Experimental Husbandry Farm (ADAS)
GRI	The Grassland Research Institute, Hurley
HDRI	The Hannah Dairy Research Institute, Ayr
HMSO	Her Majesty's Stationery Office
ICI	Imperial Chemical Industries, Ltd
IGER	The Institute of Grassland and Environmental Research (formed by the amalgamation of WPBS and AGRI)
INRA	Institut National de la Recherche Agronomique, France
MAFF	The Ministry of Agriculture, Fisheries and Food, London
MGA	The Maize Growers Association
MMB	The Milk Marketing Board
NIAB	The National Institute for Agricultural Botany, Cambridge
NIAE	The National Institute for Agricultural Engineering, Silsoe (now The Silsoe Institute)
NIRD	The National Institute for Research in Dairying, Shinfield
SAC	The Scottish Agricultural Colleges
UKASTA	The United Kingdom Agricultural Supply Trade Association
WPBS	The Welsh Plant Breeding Station, Aberystwyth

INDEX

INDEX

A
Additives, hay, 14–15
 ammonium propionate, 14
 application systems, 15
 propionic acid, 14
Additives, silage, 22–33
 AIV acid, 18, 27
 applicators, 23, 134–136
 approval scheme, 26–27
 bacterial inoculants, 29–30
 care in handling, 28
 chemical enzymes, 30–31
 effect on nutritive value, 195, 208
 formic acid, 23–25, 209
 guide to use, 32
 hydrochloric acid, 18, 27
 inorganic (mineral acids), 27–29
 molasses, 22–23
 organic acids, 28
  with formalin, 28–29
 sodium acrylate, 29
 sodium metabisulphite, 23, 29
 sodium sulphite, 29
 sugars, 29
 sulphuric acid, 18, 27–28
Agricultural Development and Advisory Service (ADAS), 26, 34, 224
 Feed Evaluation Unit, 43
Alkali, conservation of forages, 33–35
 straw, 167–171, 200, 224
 whole-crop cereals, 34, 81–85, 199
Ammonia for conservation, 34–35
Ammonia-treated straw, 167–171
 D-value, 167
 for buffer feeding, 223–224
Animal conservation and diversity, 88
Arable Area Payment Scheme, 77, 83, 218

B
Bacterial action, hay, 11
Bale-grouping devices, 109
Baler output, 110
Balers
 large rectangular, 114, 118
 round, 115–117
 standard, 108–110
Bales, 107
 collecting, 111–114
 handling in groups, 111–114
 large rectangular, 114, 118–120
 loading into store, 114
 round, 115–117
  suitability for haymaking, 116–118
 special-purpose carriers, 112
 standard, 107–110
  handling, 110–111
 see also Big bales: Rectangular bales, large: Round bales, large
Baling, general operation, 109
Barley-beef system, 1, 206–207
Barley supplement feeding
 beef cattle, 44, 216
Barn-dried hay, 12, 120
 moisture limits, 8
 nutritive value, 13
Barn hay-drying, 120
Beef cattle, silage feeding, 44, 207–209
Beef production, conserved forages, 206–212
 barley beef system, 1, 206–207

Beef production (*contd.*)
big-bale silage, 207–208
cereal and energy supplements, 209
feeding hay, 207–208
importance of forage quality, 208
protein supplements, 210–212
Big bales
for beef production, 207–208
as buffer feeding, 223
feeding from, 179–182
for sheep production, 214
silage, 150–156
see also Bales: Rectangular bales, large: Round bales, large
Block cutters, silage, 178–179
BSE crisis, 219
Buffer feeding, use of silage, 4, 223
Bunker silos *see* Silos
Butterfat levels and conserved forage, 203–204

C
Carbohydrate oxidisation after cutting, 7
Cereals, supplement feeding
beef cattle, 44, 216
Cereals, wholecrop, *see* Wholecrop cereals
Chemical enzymes as silage additives, 30–31
Chop length, 51, 122, 126, 144
Clover, red
digestibility, 63
silage and milk production, 196
Clover, white
bicropping, 85
digestibility and yield, 63, 66
soil N supply, 70
Cocksfoot, digestibility, 62–63, 65–66
Common Agricultural Policy, 5, 219, 225
Compensation for less intensive farming, 88–89
Complete diet feeding systems, 183, 224–225
Computer-generated feeding, 58
Concentrate feeding
effects on forage digestibility and intake, 53
to beef cattle, 209–212
to dairy cattle, 200–206
Conditioning
combined with mowing, 91–92
see also Mowing
drying rate, 105–106
losses, 106
Controlled feeding, shift towards, 224
Costs of forage conservation, 221
Countryside Stewardship Scheme, 88
Crimping, 98–100
Crop conditioning *see* Mowing and conditioning
Crop loss and drying rate, 105–106
Crops for conservation, 59–89
dry matter production, 59
forage maize, 76–81
for grass drying, 59
grass for grazing, 59
for hay, 74–75
for high-temperature drying, 87–88
kale and crop by-products for silage, 86–87
for silage, 75–76
sward management, yield and quality, 66–69
wholecrop cereals, 81–86
Crushing, 98–100
Curing hay in bales, 12
Cutter-bar mowers, 92
Cutting
dates, and digestibility, 39–42
first cut
D-value and yield, 62–66
effect on regrowth, 67–68
of grazing land, 61–66
methods, *see* Mowing and conditioning: and Mower types
chop length, 129

D
D-value *see* Digestibility
Dairy cows, conserved forage feeding
concentrate supplementation, 200–203
dry cow feeding, 205–206

milk production, 191–203
see also Milk production
milk quality, 203–205
experiments with high and low D-value silage, 191–196
grazing and milk output, 68–69
Department of Agriculture, Northern Ireland, 26, 209, 211, 214, 221
Digestibility, 37–45
different forage species, 62–66
of conserved forages, 42–45
effect of different methods, 42–45
and date of cutting, 62–66
D-value, 37–38, 39–40
of different forage varieties, 39–40, 62–66
and early first cut, 67–68
effect of concentrate supplement, 201
effect of cutting dates, 39–40, 62–66
fibre digestion rate, 53
of forage at cutting, 39–40, 62–66
typical changes, 40–42
of hay, 44
metabolisable energy (ME), 37–38
of silage, 43
stage of crop maturity, 39–42
of straw, 164, 165, 167
and weight gains, 52, 208
of wholecrop cereal silage, 83
Diseases of forage species, 74
Dorset Wedge system, method of filling silo, 140–141
Double-chop forage harvester, 127
Dried grass
crops for high temperature drying, 87–88
feed intake, 51–52
protein value, 47, 202
Dry cow feeding, 205–206
Dry matter content
before ensilage, 20–22, 123–125
effect on silage intake, 50–51
effect on silage nutritive value, 49–51, 194, 208
of silage maize, 77, 197
of whole-crop cereals, 82–83, 84

E

Effluent loss from silage, 159–162
feeding to cattle, 162
Energy supplements and weight gains, beef cattle, 209–210
Ensilage *see* Silage
Environmental conservation, 226
Environmentally Sensitive Areas, 88
Environmentally sensitive farming, 88
Enzymes, chemical, as silage additives, 30–31
European Union, 1
dried green crop production, 16–17

F

Feed passages and feed troughs, feeding from, 184–187
Feed value of forages, 36–58
conserved forages fed in mixed rations, 52–56
differences between forages, 48–49
digestibility, 37–45
see also Digestibility
effects of particle size, 52
feed intake, 48–52
fibre in ruminant feeding, 56–57
see also Fibre
intake of silage, 49–51
dry matter content, 50–51
effect of pH, 50
protein value, 45–47
of silage, 43–45, 49–51
Feeding
conserved forages, 189–217
controlled, shift towards, 224–225
intake of silage, *ad libitum*, 50, 55–56, 57
Feeding methods, 172–188
mechanised, 177–188
bunker and clamp silos, 177–187
tower silos, 187–188
see also Big bales: Feed passages and troughs: Hay: Mixer-feeder wagons: Silage: Straw: Tower silos
Feeding system, storage, 225
Fermentable metabolisable energy, 47, 198

Fermentation
bacterial inoculants, 29–30
ensilage process, 17–20
restricted, silage, 163
Fertiliser use on grass and forage crops, 69–74
Fibre
and forage digestibility, 53
and milk-fat percentage, 55, 204
in ruminant feeding, 56–57
Finger-bar mowers, 92
Fingerless double-knife mower, 92
Fishmeal supplement
beef cattle, 211
dairy cattle, 202
Flail harvester, 92, 126
Flail mower, 93, 127
Flail mowers for conditioning, 99–100
Forage boxes, feeding from, 182–183
Forage conservation
in farming systems, 218–226
livestock industry uncertainties, 218–220
increasing costs, 220–221
principles, 7–35
Forage crops, fertiliser, 69–74
Forage harvesters, 126–130
Forage harvesting systems, 136–138
Forage quality, importance, 208–209
Forages, conserved
in beef production, 206–208
in dairy cow feeding, 190–206
see also Dairy cows: Milk production
in dry cow feeding, 205–206
feed value, 36–58, 189–217
for sheep, 212–216
Formalin silage additive, 28–29, 163
Formic acid and formalin silage additive, 196–197
silage additive, 23–25
Fuel for driers, cost, 16

G

Grass, drying rate, 95, 105–106, 123–125
Grass for conservation, 59–66
Grass, fertiliser use, 69–74
Grass, high-temperature drying, 15–17
crops for drying, 87–88
EU production, 16–17
Grass for grazing, 59–66, 221–222
integration with cutting, 61–66
rotational, 60–61
various grasses, 62–66
Grass management
effect of cutting on regrowth, 67
yield and quality, 66–69
Grass silage
effect on milk quality, 204–205
mixed with maize silage, 198
intake and milk production, 198
Grassland Research Institute, 34, 61
Grazed grass, extending the season, 221–222

H

Harvester, forage, types, 126–130
Harvesting, date and D-value, 62–66
Harvesting systems for haymaking, 107–120
Harvesting systems for silage-making, 136–138
Hay, barn-dried, 109–110
barn-drying, 12–14, 121
digestibility, 13
moisture content for storage, 8–10
Hay, field dried, 8–12, 107–120
baling, 107–120
effect of weather, 9–12
methods of feeding, 172–173
moisture content for storage, 8–10
mowing and conditioning, 90–106
Haymaking, 8–15, 107–120
additives, 14–15
bales and balers, 107–120
see also Balers: Bales
cutting and conditioning, 105
mould prevention, 14
and storage moisture content, 7, 8–9
field exposure time, 7, 8–9
mature/immature crop, 8–9
tedding, 11, 96
Herbicides, 79–80
High-temperature drying *see* Grass drying
Hydraulic shear grab, 178–179

Hydrochloric acid as silage additive, 27

I

Indoor bunker silos, effluent loss, 161
Indoor walled silos, 141–146
Institute for Grassland and Environmental Research (IGER), 30, 86, 192, 201, 204

K

Kale and crop by-products for silage, 86–87

L

Lactic acid as silage additive, 28
Lactobacilli, 19, 30
Lactobacillus plantarum, 30
Legumes, D-values, 41, 63
Lucerne digestibility, 63

M

Magnesium, fertiliser, 72
Maize, forage, 76–81
  areas suitable, 78–81
  ensiling, 158–159
  harvesting for silage, 130–132
  for silage, 76–81
Maize gluten, effect on weight gain, 211
Maize Growers Association, 34, 79, 83
Maize silage
  additives, 158–159
  dry matter, 197–198
  method of feeding, 176, 225
  mixed with grass silage, 198
    effect on milk quality, 204
    intake and milk production, 198–199
Mechanised feeding methods, 177–188
Metabolisable energy of conserved forages, 37–38, 44
Milk composition, 203–205
Milk Marketing Board, 192, 201, 203, 204
Milk prices, 218–219, 220
Milk production, 190–205
  grass/maize silage feeding, 198
  high or low D-value silage feeding, 191–196
  and milk prices, 218– 220
  wholecrop cereal silage, 199
Milk quality, and conserved forages, 203–205
Milk quotas
  effect on dairy cow feeding, 4, 191, 220–221
  and forage maize, 77–78
Mineral content of conserved forages, 57–58
Mineral fertiliser requirements of forage crops, 72–74, 78–81
Mixed rations, conserved forages fed in, 52–56, 200–203, 209–212
Mixer-feeder wagons
    complete diet feeding systems, 224
    feeding from, 183–184
Moisture content, field exposure times, 7, 8
Molasses, silage additive, 22–23, 29
Mowers, types, 92–95
Mowing and conditioning, 90–106
  avoiding double-cutting, 94–95
  basic requirements, 91–92
  combined equipment, 101
    operating, 101–105
  conditioning equipment, 98–101
  crop conditioning, 98
  crushing and crimping, 98–100
  mowing equipment, 92–95
  treatment after mowing, 95–98

N

National Institute of Agricultural Botany, 61
Near Infrared Reflectance spectroscopy, 38–39, 43–44
Nitrogen fertilisers
  and forage carbohydrate content, 75–76
  use on forage maize, 79–81
  use on grassland, 69–74
  use on wholecrop cereals, 84–85
Nutritive value
  of conserved forage, 36–56

Nutritive value (*contd.*)
of hay, 43
high-temperature dried grass, 15–16, 43
silage, 43, 49

O

Outdoor bunker silos, 147–150
effluent loss, 161

P

Pediococcus acidilactici, 30
Plant conservation and diversity, 88–89
Plastic sealing of silos
indoor silos, 144–146
large rectangular bales, 154–166
outdoor silos, 147–150
round bales, 151–154
weighting the silo surface, 144, 149
Pre-wilting the crop *see* Wilting
Precision-chop (metered) forage harvesters, 127
Propionic acid
silage additive, 28
hay additive, 14
Protein
breakdown after cutting, 7
metabolism, 45–47
rumen degradable, 45–47, 55
Protein content of supplement, effect on milk quality, 204
Protein supplements
for beef cattle, 210–212
for dairy cattle, 200–203
Protein value of conserved forages, 45–47, 201
crude protein, 45, 46–47
efficiency of use, 47

R

Raking, 96–98
Reciprocating finger-bar mowers, 92
Rectangular bales, large, 118–120, 154–156
feeding methods, 173
unwrapping, 173
Rotary disc mower, 93
Rotary drum mower, 93
Round bales, large, 115–117, 150–154
automatic wrapping, 152–154
feeding from, 181–182
feeding methods, 173
unwrapping, 173
pre-chopping, 154
storage, 154
Rowett Research Institute, 206
Ryegrass, digestibility and yield, 39, 63, 64
Ryegrass silage
effect on milk quality, 204
and milk production, 196

S

Sainfoin, digestibility, 49, 63
Scottish Agricultural Colleges, 26, 223
Self-feeding of silage, 175–177
electric fencer, 176–177
short-chopped forage, 176–177
Self-loading forage wagons, 138
Shear grab, hydraulic, 178–179
Sheep, conserved forages, 212–216
effect of hay, straw and silage, 212–216
D-value, 214–216
during pregnancy, 214–215
supplementary feeding, 215–216
Silage
as buffer feeding, 223
care during storage, 162–163
crops for, 75–76
effects of long/short cutting, 51
feeding value, *see* Feeding value
intake, 49–51
moisture content limits, 8
principles of ensilage, 17–33
quantities conserved, 1960–93, 3
with restricted fermentation, 163
Silage Additive Scheme, 26, 134
Silage feeding methods, 173–188
big bales, 179–182
feed passages and feed troughs, 184–187
forage boxes, 182–183
self-feeding, 175–177
unwrapping bales for feeding, 173, 174

Silage making, 17–33, 121–163
additives, 18–19, 22–33, 134–136
see also Additives
avoiding soil contamination, 138–140
dry-matter content: the 'wilting' debate, 123–125
field decisions, 122
filling the silo, *see* Silos
forage harvesting systems, 136–138
harvesting the crop, 123–140
harvesting forage maize for silage, 130–132
microbial activity, 19
wilting, *see* Wilting
Silo unloaders, 178–179
Silos
clamp, 148–150
Dorset Wedge, 140–150
filling, 140–163
equipment, 146
procedure, 142–146
indoor walled, 141–146
outdoor bunker, 147–150
sealing, 141, 144, 147, 149
tower, 156–158
Silsoe Working Party, 81
Sites of Special Scientific Interest, 88
Sodium acrylate as silage additive, 29
Sodium bicarbonate, effect of including with concentrate, 56
Sodium hydroxide-treated straw as animal feed, 169–171, 200
high sodium content, 170–171
Sodium hydroxide treatment of straw, 164–165, 169–171
Sodium metabisulphite as silage additive, 23, 29
Sodium sulphite as silage additive, 29
Soil contamination of silage, avoiding, 138–140
Soyabean, effect on weight gain, 207, 211
Storage feeding system, 225
Straw as animal feed, 164–171
ammonia treatment, 166–169
availability, 165
collecting and storing, 165–166
digestibility (D-value), 164, 165, 167
sodium hydroxide treatment, 169–171
untreated, 166–167
Straw feeding methods, 172–173
chopping, 173
feed mixer wagons, 169–170
unwrapping large rectangular bales, 173
Sugar-beet tops for silage, 86
Sugars as silage additives, 29
Sulphur deficiency in forages, 72–73
Sulphuric acid as silage additive, 27–28
care in handling, 28
Sward management, yield and quality, 66–69

T
Tedders and tedding, 11–12, 96–98
Timothy grass, digestibility, 65
Tower silos, 21, 156–158
mechanised feeding from, 187–188
Trace elements, supply and uptake, 73

U
UK Agricultural Supply Trade Association (UKASTA), 26
/Advisory Services Silage Additive Approval Scheme, 26
Urea for forage conservation, 34–35, 82–85, 199

V
Vegetable wastes for silage, 86–87

W
Weather conditions and haymaking, 9–12
Weed control, 79
Weight gains
cereal supplements, beef cattle, 209
energy supplements, beef cattle, 209
importance of forage quality, 44, 208–209

Weight gains (*contd.*)
and protein supplements for beef cattle, 211–212
Wholecrop cereal silage, 81–86, 132–134
advantages, 85
for buffer feeding, 223–224
conventional and bi-cropping, 85
digestibility, 83
dry matter, 82–83, 84, 197–198
ensiling, 33–35, 159
growers' guides to maturity, 34–35
later maturity, effect, 33–34
nitrogen input, 84–85
research into alkali treatment, 34
urea conservation, 34
effect on milk production, 198–199
Wholecrop Cereals Group, 83
Wildlife and Countryside Act, 88
Wilting for ensilage, 20–22, 50
crop conditioning, 95–105
effect on weight gains, 208–209
and dry-matter content, 123–125
effect on feed value, 47, 50–51
effect on milk production, 194–196
field, 21
practicability, 32–33
Wye College, 194, 219

ABOUT THE AUTHORS

Frank Raymond worked at the Grassland Research Institute at Hurley, where his research was particularly concerned with forage nutritive value and methods of forage conservation. In 1972 he joined the Ministry of Agriculture, Fisheries and Food in London and served as Chief Scientist there from 1981 to 1982. Since then he has advised the European Union, FAO and a number of national Ministries on agricultural research, planning and programmes; he has been much involved with the work of RURAL, the Society for the Responsible Use of Resources in Agriculture and on the Land, and has also been a member of the Policy Committee of the Council for the Protection of Rural England. He has been President of both the British Grassland Society and the British Society of Animal Production.

Richard Waltham, who has spent his farming life in North Dorset, became interested in forage conservation in the 1950s, when silage-making expanded on the farm. Better control of both fermentation and loss in storage led to the Dorset Wedge system – now accepted and adopted by farmers generally. He has been President of the British Grassland Society and Chairman of the Royal Smithfield Club.